Toward Achieving a Carbon-Neutral Society: Beneficiation and Extractive Metallurgy for Producing Critical Metals from Ores/Wastes

Toward Achieving a Carbon-Neutral Society: Beneficiation and Extractive Metallurgy for Producing Critical Metals from Ores/Wastes

Guest Editors

Ilhwan Park

Sanghee Jeon

MDPI

Basel • Beijing • Wuhan • Barcelona • Belgrade • Novi Sad • Cluj • Manchester

Guest Editors
Ilhwan Park
Division of Sustainable Resources Engineering
Hokkaido University
Sapporo
Japan

Sanghee Jeon
Graduate School of Engineering and Resource Science
Akita University
Akita
Japan

Editorial Office
MDPI AG
Grosspeteranlage 5
4052 Basel, Switzerland

This is a reprint of the Special Issue, published open access by the journal *Metals* (ISSN 2075-4701), freely accessible at: https://www.mdpi.com/journal/metals/special_issues/NK5YZ7RD37.

For citation purposes, cite each article independently as indicated on the article page online and as indicated below:

Lastname, A.A.; Lastname, B.B. Article Title. *Journal Name* **Year**, *Volume Number*, Page Range.

ISBN 978-3-7258-6692-2 (Hbk)
ISBN 978-3-7258-6693-9 (PDF)
https://doi.org/10.3390/books978-3-7258-6693-9

Contents

About the Editors

Ilhwan Park

Ilhwan Park is an associate professor at the Division of Sustainable Resources Engineering, Faculty of Engineering, Hokkaido University, Japan. He obtained his Ph.D. in the field of Sustainable Resources Engineering from Hokkaido University, Japan, in 2019. After his graduation, he was appointed as an assistant professor at the same university before being promoted to the position of associate professor in 2023. Dr. Park has acted as Editor of *Resources Recycling* (*The Korean Institute of Resources Recycling, KIRR*), Associate Editor of *Frontiers in Environmental Science*, and Section Board Member of *Metals* (MDPI) and *Minerals* (MDPI). His research interests cover the recovery of critical metals/minerals from primary and secondary resources by mineral processing (e.g., flotation, magnetic separation, gravity separation) and hydrometallurgy (e.g., leaching, solvent extraction, cementation). In addition, Dr. Park has been engaged in the development of sustainable techniques for the prevention/control of acid mine/rock drainage (AMD/ARD) and the remediation of contaminated soils and water. From his work, he has published 75 papers in leading international journals in the fields of mineral processing, extractive metallurgy, and environmental science.

Sanghee Jeon

Sanghee Jeon has been a full associate professor of the Faculty of International Resource Science, Akita University, which is one of the strongest universities in mining, mineral processing, and metallurgy in Japan. Before joining Akita University in 2023, the researcher earned a PhD at Hokkaido University in the Department of Sustainable Resource Engineering in 2019. Her research interests include mineral processing, resource recycling, and hydrometallurgy, especially specializing in the recovery of precious and critical metals from both primary and secondary resources employing sustainable and eco-friendly techniques. The researcher has recently focused on the development of novel recovery techniques for precious metals from eco-friendly solvents such as thiosulfate, including the utilization of tailings, recovery of critical metals from end-of-life Li-ion batteries, and rare earth element (REE) ore treatment, as well as deep-sea mining (i.e., hydrothermal ore treatment). The researcher received 18 scientific awards in total, including the 2018 IMPC Young Author Award, the Otsuka Award in Hokkaido University, the Faculty Evaluation Excellence Award in Akita University, and so on. And she published 78 international scientific articles published in journals indexed in the Web of Science and Scopus (total citations: 3191; i10-index of 48 in Google Scholar).

metals

MDPI

Editorial

Toward Achieving a Carbon-Neutral Society: Beneficiation and Extractive Metallurgy for Producing Critical Metals from Ores/Wastes

Ilhwan Park [1,*] and Sanghee Jeon [2]

[1] Division of Sustainable Resources Engineering, Faculty of Engineering, Hokkaido University, Sapporo 060-8628, Japan
[2] Department of Earth Resource Engineering and Environmental Science, Faculty of International Resource Science, Akita University, Akita 010-0865, Japan; jeon@gipc.akita-u.ac.jp
* Correspondence: i-park@eng.hokudai.ac.jp

1. Introduction and Scope

The global commitment to achieving a carbon-neutral society has accelerated the transition toward renewable energy, electric mobility, and advanced electronic systems [1]. This transition, however, has led to an unprecedented increase in the demand for a wide range of critical metals, including copper (Cu), lithium (Li), nickel (Ni), cobalt (Co), and rare-earth elements (REEs) [2]. According to the International Energy Agency (IEA), the demand for key energy-transition minerals in 2040 is projected to increase by approximately 1.2 times for Cu, 4.5 times for Li, 1.7 times for Ni, 1.5 times for Co, and 1.6 times for REEs compared with their respective demand levels in 2024 [3].

To meet the future demand for these metals, technological advancements in beneficiation and extractive metallurgical processes for both primary ores and secondary resources—such as mine tailings, metallurgical slags, industrial residues, electronic waste, and wastewaters—are essential. Beneficiation and extractive metallurgy play a central role in enabling such low-carbon metal supply chains by integrating physical separation, chemical extraction, and environmentally benign process design.

This Special Issue, entitled "Toward Achieving a Carbon-Neutral Society: Beneficiation and Extractive Metallurgy for Producing Critical Metals from Ores/Wastes", aims to highlight recent scientific and technological advances in the sustainable production of critical metals from both primary and secondary resources. The scope covers a wide range of processes, including flotation, magnetic and gravity separation, thermal reduction, leaching, solvent extraction, adsorption, and emerging hybrid and low-energy technologies.

2. Contributions

This Special Issue comprises nine articles that collectively demonstrate how modern beneficiation and extractive metallurgy can contribute to carbon-neutral and circular production of metals, including Li, gold (Au), Cu, Ni, Co, and REEs.

Eom et al. (Contribution 1) provided a comprehensive review of Li extraction from primary hard-rock resources, with a particular focus on mechanochemical treatment as a promising strategy for improving Li recovery. The review highlights recent developments in mechanochemical activation as an energy-efficient pretreatment to enhance Li leachability from spodumene- and lepidolite-type minerals.

Mhandu et al. (Contribution 2) studied a thiosulfate-based green hydrometallurgical process as an environmentally benign alternative to conventional cyanide-based gold extraction. It has been reported that the efficiency of thiosulfate leaching of gold from sulfidic refractory ores is low due to the passivation of the gold surface and/or the decomposition of thiosulfate [4–6]. The mechanistic investigations of Mhandu et al. (Contribution 2) to elucidate the detrimental effects of arsenopyrite/pyrite during thiosulfate leaching of gold revealed that thiosulfate decomposition on the arsenopyrite/pyrite surface—rather than passivation of the gold surface—is the primary cause of the drastic decrease in gold extraction efficiency. Furthermore, Mhandu and coworkers (Contribution 2) proposed a new flowsheet incorporating a pre-oxidation step using cupric ammine complexes prior to thiosulfate leaching, which significantly improved gold extraction efficiency from 10% to 79%.

Although thiosulfate is a promising lixiviant, recovering gold from thiosulfate media remains challenging. Cementation employing aluminum (Al) as the electron donor and activated carbon (AC) as the electron mediator has been reported to enable effective recovery of gold ions from thiosulfate media [7]; however, Cu ions present as tetraammine complexes—acting as the oxidant for gold—were simultaneously recovered, resulting in increased reagent consumption upon reuse of the lixiviant. Zoleta et al. (Contribution 3) demonstrated that substituting AC with iron oxides, such as hematite (Fe_2O_3) and magnetite (Fe_3O_4), enables the Al-based cementation process to become selective for gold ions, contributing to the practical implementation of cyanide-free gold processing.

Fang et al. (Contribution 4) investigated the applicability of microwave treatment for enhancing the grinding and flotation performance of copper–nickel sulfide ore. It was found that the grindability of the ore improved with increasing microwave treatment duration, as evidenced by reductions in D_{80} and relative work index (RWI) values, along with an increase in the degree of liberation. However, excessively prolonged microwave treatment reduced flotation efficiency due to surface oxidation of the minerals. Under optimal conditions, microwave treatment was demonstrated to effectively improve both grinding and flotation performance.

Simonič et al. (Contribution 5) investigated the potential of bio-based secondary materials (i.e., torrefied wood-waste biomass) for the adsorption of Cu(II) and Ni(II) ions, demonstrating that thermally treated biomass can serve as an efficient and low-cost sorbent for metal removal and recovery from aqueous systems. In addition, the torrefied biomass exhibited enhanced hydrophobicity and improved energy properties, highlighting its potential as an improved solid fuel.

Gong et al. (Contribution 6) studied the reprocessing of Co-bearing Cu slag via carbothermic reduction followed by magnetic separation to recover Co, which is an indispensable metal for the energy transition, particularly for the production of Li-ion batteries. Carbothermic reduction treatment altered the distribution of Co, which was mainly present in iron silicate minerals in the pristine slag but predominantly existed as cobalt–iron alloy in the roasted slag. As a result, the magnetic separation concentrate exhibited an increase in Co grade from ~1.3% to 4.0% with a recovery of 94.2%, demonstrating that carbothermic reduction followed by magnetic separation is a promising approach for transforming slags into secondary cobalt resources.

Similarly, Lim et al. (Contribution 7) emphasized the importance of reprocessing non-ferrous slags (e.g., tin, copper, nickel, vanadium, and titanium slags) for the recovery of critical and strategic metals. The review article by Lim and coworkers (Contribution 7) systematically analyzes pyrometallurgical and hydrometallurgical routes for metal recovery from industrial slags and highlights their role as long-term urban and industrial metal reservoirs.

Rare-earth elements are also critical for transitioning toward a low-carbon future. Jürjo et al. (Contribution 8) focus on the leaching and extraction of REEs and associated

https://doi.org/10.3390/met16010113

rare metals from Estonian graptolite-argillite and phosphorite ores, demonstrating the feasibility of valorizing non-traditional geological resources as alternative REE sources.

Finally, Litvinova et al. (Contribution 9) provided fundamental insights into REE behavior through a thermodynamic study of rare-earth ion associations in the Ln^{3+}—CO_3^{2-}—H_2O system, elucidating carbonate complexation and ion-pair formation mechanisms relevant to alkaline and carbonate-based extraction processes.

Together, these contributions span a wide spectrum of resources—ranging from primary ores to slags, industrial wastes, and biomass—and employ diverse physical, chemical, and hybrid process routes, clearly illustrating the multifaceted nature of modern sustainable metallurgy.

3. Conclusions and Outlook

This Special Issue has presented recent progress in beneficiation and extractive metallurgical technologies for the sustainable production of critical metals from both primary ores and secondary resources. The collected papers clearly demonstrate that advanced physical separation, innovative hydrometallurgical and pyrometallurgical processes, and bio-based and low-carbon approaches can significantly contribute to resource efficiency and carbon-neutral metal production. Despite the significant progress made in this field, numerous challenges remain unresolved. In this context, this Special Issue is intended to serve as a springboard for future scientific discussion and debate on emerging and challenging topics related to the sustainable supply of critical metals.

Funding: This research received no external funding.

Conflicts of Interest: The authors declare no conflicts of interest.

List of Contributions:

1. Eom, Y.; Dyer, L.; Nikoloski, A.N.; Alorro, R.D. Mechanochemical Treatment for the Extraction of Lithium from Hard Rock Minerals: A Comprehensive Review. *Metals* **2024**, *14*, 1260.
2. Mhandu, T.J.; Park, I.; Jeon, S.; Hamatsu, S.; Elakneswaran, Y.; Ito, M.; Hiroyoshi, N. A Pretreatment of Refractory Gold Ores Containing Sulfide Minerals to Improve Gold Leaching by Ammonium Thiosulfate: A Model Experiment Using Gold Powder and Arsenic-Bearing Sulfide Minerals. *Metals* **2023**, *13*, 1357.
3. Zoleta, J.; Jeon, S.; Kuze, A.; Okada, N.; Park, I.; Ito, M.; Elakneswaran, Y.; Hiroyoshi, N. Selective Cementation of Gold using an Iron Oxide and Zero-Valent Aluminum Galvanic System from Gold-Copper Ammoniacal Thiosulfate Solutions. *Metals* **2023**, *13*, 1289.
4. Fang, X.; Peng, Z.; Yin, T.; Rao, M.; Li, G. Microwave Treatment of Copper–Nickel Sulfide Ore for Promotion of Grinding and Flotation. *Metals* **2024**, *14*, 565.
5. Simonič, M.; Goričanec, D.; Petrovič, A.; Šilić, I.; Urbancl, D. Cu(II) and Ni(II) Adsorption on Torrefied Wood Waste Biomass. *Metals* **2025**, *15*, 304.
6. Gong, J.; Pan, J.; Zhao, J.; Zhang, Q.; Hao, G.; Liu, Y.; Yu, H. Selective Recovery and Enrichment of Cobalt from Cobalt-Containing Slag by Carbothermic Reduction. *Metals* **2025**, *15*, 622.
7. Lim, B.; Aylmore, M.; Alorro, R.D. Technospheric Mining of Critical and Strategic Metals from Non-Ferrous Slags. *Metals* **2024**, *14*, 804.
8. Jürjo, S.; Siinor, L.; Siimenson, C.; Oll, O.; Lust, E. Extraction of REEs and Other Elements from Estonian Graptolite-Argillite and Phosphorite Mineral Acid Solutions. *Metals* **2025**, *15*, 608.
9. Litvinova, T.; Gerasev, S.; Sergeev, V.; Lidanovskiy, E. Rare Earth Metal Ion-Associates in Ln^{3+}—CO_3^{2-}—H_2O system. *Metals* **2025**, *15*, 239.

https://doi.org/10.3390/met16010113

References

1. International Renewable Energy Agency. *Untapped Potential for Climate Action: Renewable Energy in Nationally Determined Contributions*; IRENA: Abu Dhabi, United Arab Emirates, 2017.
2. Hund, K.L.; La Porta, D.; Fabregas, T.P.; Laing, T.; Drexhage, J.R. *Minerals for Climate Action: The Mineral Intensity of the Clean Energy Transition*; World Bank Group: Washington, DC, USA, 2020.
3. International Energy Agency. *Global Critical Minerals Outlook 2025*; IEA: Paris, France, 2025.
4. Feng, D.; van Deventer, J.S.J. Ammoniacal thiosulfate leaching of gold in the presence of pyrite. *Hydrometallurgy* **2006**, *82*, 126–132. [CrossRef]
5. Xu, B.; Yang, Y.; Li, Q.; Jiang, T.; Zhang, X.; Li, G. Effect of common associated sulfide minerals on thiosulfate leaching of gold and the role of humic acid additive. *Hydrometallurgy* **2017**, *171*, 44–52. [CrossRef]
6. Liu, X.; Xu, B.; Min, X.; Li, Q.; Yang, Y.; Jiang, T.; He, Y.; Zhang, X. Effect of Pyrite on Thiosulfate Leaching of Gold and the Role of Ammonium Alcohol Polyvinyl Phosphate (AAPP). *Metals* **2017**, *7*, 278. [CrossRef]
7. Jeon, S.; Bright, S.; Park, I.; Tabelin, C.B.; Ito, M.; Hiroyoshi, N. A simple and efficient recovery technique for gold ions from ammonium thiosulfate medium by galvanic interactions of zero-valent aluminum and activated carbon: A parametric and mechanistic study of cementation. *Hydrometallurgy* **2022**, *208*, 105815. [CrossRef]

https://doi.org/10.3390/met16010113

metals

MDPI

Article

Selective Recovery and Enrichment of Cobalt from Cobalt-Containing Slag by Carbothermal Reduction

Jiachen Gong [1,2], Jian Pan [1,*], Jingfu Zhao [2], Qian Zhang [1,*], Guansheng Hao [2], Yan Liu [2] and Helei Yu [2]

1 School of Minerals Processing and Bioengineering, Central South University, Changsha 410083, China; 235604053@csu.edu.cn
2 CNMC Shenyang Research Institute of Nonferrous Metals Co., Ltd., Shenyang 110141, China; 15840325215@126.com (J.Z.); haozijun777@163.com (G.H.); 15566531662@163.com (Y.L.); yuhelei1986@163.com (H.Y.)
* Correspondence: pjcsu@csu.edu.cn (J.P.); vqianzhang@csu.edu.cn (Q.Z.)

Abstract: Cobalt ore resources are relatively scarce; thus, the recycling of cobalt-containing slag is highly significant in the economy and society. In this study, the effects of reduction temperature, the reduction agent ratio, reduction time, and particle size on the grade and recovery rate of cobalt in a concentrate were systematically investigated during the carbothermal reduction of cobalt-containing slag. The results revealed that the grades of cobalt, iron, and copper in the concentrate after magnetic separation were 4.02%, 2.48%, and 81.33%, respectively, and the recoveries were 94.17%, 74.80%, and 53.27%, respectively, under the reduction temperature of 1150 °C, the reduction agent ratio of 40%, the reduction time of 2 h, and the particle size of −3.0 mm. Furthermore, through static reduction roasting in a muffle furnace and dynamic reduction roasting in a rotary kiln followed by magnetic separation, a stable cobalt grade, high selective recovery, and effective enrichment were achieved under optimal conditions.

Keywords: cobalt-containing slag; carbothermal reduction; selective recovery; cobalt grade

1. Introduction

Cobalt is an important, strategic non-ferrous metal widely used in fields such as battery materials, high-temperature alloys, hard alloys, magnetic materials, and catalyst [1,2]. With the increasing global demand for clean energy and electric vehicles, the status and future prospects of cobalt resources have attracted widespread attention [3,4].

From the perspective of cobalt ore resources, pure cobalt ores are rare, and cobalt is almost entirely present as a by-product in associated ores, with a very low concentration at 25 ppm in the Earth's crust [5,6]. The main types of cobalt ores include the following: sandstone-type copper ores, lateritic nickel ores, and magmatic nickel–copper sulfide deposits [7]. Currently, the majority of the globally proven cobalt resources are contained in nickel laterite deposits, with the remainder mostly found in nickel–copper sulfide deposits [8,9]. These deposits are primarily located in countries such as Australia, Canada, and Russia [10]. Additionally, cobalt resources in the Democratic Republic of Congo (DRC) and the Republic of Zambia are mainly found in sedimentary copper deposits [11]. The globally proven terrestrial cobalt resources amount to approximately 25 million tonnes, with reserves of 7.2 million tonnes. During the smelting of these ores, cobalt is typically extracted as a by-product. In China, cobalt ore is distributed over a wide area [12], but cobalt resources remain relatively scarce, manifested in small reserves, low ore grades, a high proportion of low-grade ores, and a large number of associated ores [13–15]. This results in a certain degree of dependence on external sources for China's cobalt supply [16]. Meanwhile, driven by the development of the domestic lithium battery and alloy smelting industries, China's consumption of cobalt products has experienced rapid growth, increasing from 3800 tons in 2000 to 119,000 tons in 2022 [17,18]. Therefore, the recovery of cobalt from secondary resources, such as cobalt-containing slag, has become an urgent issue [19,20].

Cobalt-containing slag primarily originates from the smelting processes of cobalt and nickel ores, including high-temperature smelting, leaching, and electrolysis [21,22]. During the smelting process, the extraction efficiency of cobalt is often low, resulting in the generation of large amounts of cobalt-bearing waste slag [23]. Depending on the specific smelting process, the composition of the slag varies, typically containing various metal oxides (such as CoO, NiO, FeO, SiO_2, etc.) and other inorganic substances [24]. Recent studies have shown that the proper treatment and utilization of cobalt-containing slag can not only reduce waste accumulation but also enable resource recovery [25]. Currently, the treatment methods for cobalt-containing slag are classified into hydrometallurgical and pyrometallurgical processes [26,27]. Among them, pyrometallurgical processes are favored over hydrometallurgical ones due to their simplicity [28] and absence of extensive reagent consumption, and they remain the predominant approach for processing cobalt-containing slag [29,30]. The carbothermal reduction roasting process involves the addition of reductants to treat cobalt-containing slag at elevated temperatures. This promotes the selective reduction of cobalt and iron oxides, thereby creating favorable conditions for subsequent separation through magnetic or hydrometallurgical techniques [31,32].

This study investigates the process of selectively recovering cobalt from cobalt-containing slag via carbothermic roasting. The effects of reduction temperature, reductant ratio, reduction time, and particle size on the cobalt grade and recovery rate are examined, with both static reduction roasting in a muffle furnace and dynamic reduction roasting in a rotary kiln used for verification. Through analytical methods such as optical microscopy and scanning electron microscopy (SEM), the phase transformations during the roasting process are clarified, providing a foundational study for the industrial-scale comprehensive recovery of cobalt and other valuable metals from cobalt-containing slag.

2. Materials and Methods

2.1. Materials

The raw material of the cobalt-containing slag was obtained from the Zambia Chambishi Copper Smelting Co., Ltd., Kitwe, Zambia (CCS). The chemical compositions of the three samples were determined by ICP and XRF analyses, and the results are presented in Table 1. As listed in Table 1, the main metal elements in cobalt-containing slag are Fe, Cu, and Co, with an Fe content of 46.92~47.29%, a Cu content of 1.01~1.03%, and a Co content of 1.27~1.29%. It illustrates that Fe, Cu, and Co are uniformly distributed in the slag. The mineral content of cobalt-containing slag was statistically analyzed by optical microscope, and the results are shown in Table 2. It can be observed that the primary minerals in the cobalt-containing slag are iron silicate minerals, glassy minerals, magnetite, and copper sulfide minerals.

The XRD analysis of the cobalt-containing slag is displayed in Figure 1, where the main minerals are iron-bearing silicate minerals, magnetite, and cobalt-bearing magnetite. Cobalt-containing slag was analyzed using a scanning electron microscope (SEM) and an energy-dispersive spectrometer, and the results are shown in Figure 2. It can be seen that the phases of Fe and Co are highly overlapped and have a good correlation, indicating that there is a dispersion of Co in the ferrosilicate minerals. Si and Co also have a certain correlation, indicating that the Co distribution in the vitreous minerals may be lower.

The reduction agent used in the reduction roasting experiments of the cobalt-containing slag was anthracite. Its industrial analysis was carried out in accordance with the [33], and the results are presented in Table 3. The high fixed carbon content of the qualified anthracite ensures the efficient reduction of cobalt oxides, while the moderate volatile matter content contributes to establishing a reducing atmosphere. However, the ash content of 13.48% requires attention due to its potential impact on the melting behavior of the slag during the roasting process. These characteristics indicate that this reduction agent is suitable for the reduction roasting of cobalt-containing slag.

Table 1. Chemical composition of the cobalt-containing slag (wt. %).

Serial Number	Chemical Composition							
	Si	Al	Ca	Mg	S	Cu	Co	Fe
1	13.96	0.516	0.456	0.353	0.060	1.01	1.27	47.29
2	14.07	0.444	0.420	0.360	0.058	1.03	1.29	46.92
3	14.00	0.964	0.438	0.354	0.061	1.03	1.27	47.06

Table 2. Distribution of cobalt in cobalt-containing slag (%).

Mineral	Content	Cobalt Distribution Rate
Cobalt–nickel sulfide	0.10	2.52
Copper sulfide	1.87	0
Magnetite	5.01	0.49
Pyrrhotite	0.02	0.06
Metallic copper	0.05	0
Iron-containing silicate	77.88	96.93
Vitreous mineral	15.07	0
Cobalt–iron alloy	-	-

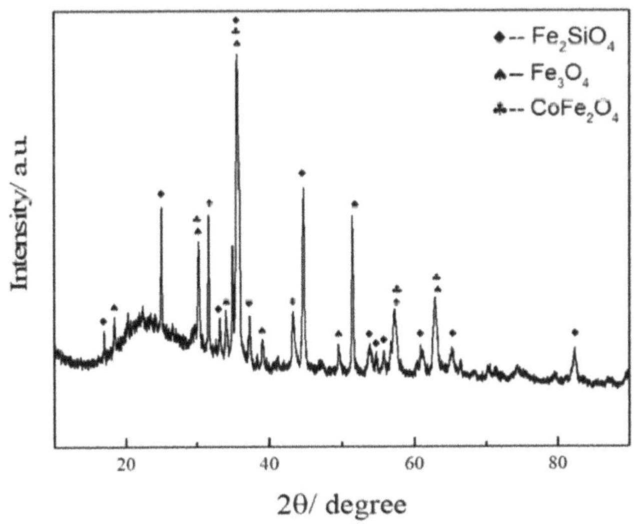

Figure 1. XRD pattern of cobalt-containing slag.

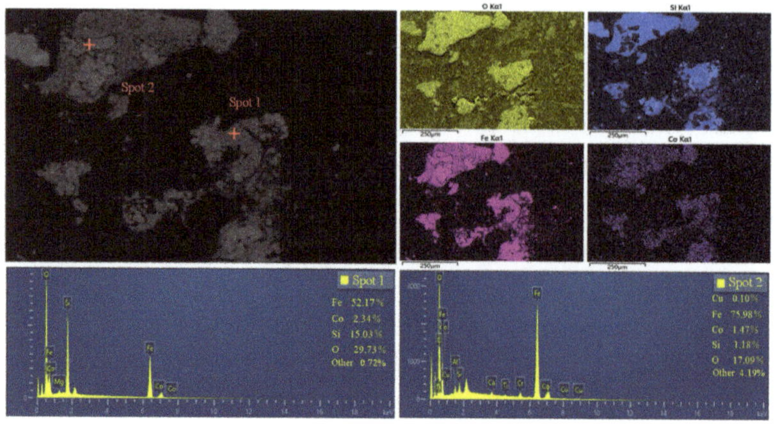

Figure 2. SEM images with EDS element mapping of the cobalt-containing slag.

Table 3. Industrial analysis results of the reduction agent.

Fixed Carbon/wt. %	Volatile Matter/wt. %	Ash Content/wt. %	Moisture Content/wt. %
67.21	17.76	13.48	1.55

2.2. Reduction Roasting Experiments and Analysis Method

2.2.1. Pretreatment of Cobalt-Containing Slag

The results of the particle size screening experiment of the cobalt-containing slag are seen in Table 4. The proportion of cobalt-containing slag with a particle size above +3 mm is 13.81%, and the proportion of cobalt-containing slag with a particle size of −3 mm to +1 mm is higher, accounting for 43.98%. Since the coarse particle size of raw cobalt-containing slag fails to meet the requirements for subsequent reduction experiments, the samples need to be pretreated. The pretreatment process of an experimental sample of cobalt-containing slag is shown in Figure 3. The cobalt-containing slag was crushed by a jaw crusher (RK/PEF 100 × 100, Wuhan Exploration Machinery Co., Ltd., Wuhan, China) and a roller crusher (RK/PG-φ200 × 125, Wuhan Exploration Machinery Co., Ltd., Wuhan, China), resulting in a particle size of −3 mm, with 80% of the cobalt-containing slag falling within this range. The cobalt-containing slag was thoroughly mixed to ensure the uniform distribution of particles of different sizes. The mixture was then split using a sample splitter, and samples were taken for further analysis.

Table 4. The results of the particle size screening experiment.

Particle Size	Weight/g	Distribution Rate/%	Positive Accumulation/%	Negative Accumulation/%
+4.00	17.67	1.39	1.39	100.00
−4.00 + 3.35	40.05	3.15	4.54	98.61
−3.35 + 3.00	117.81	9.27	13.81	95.46
−3.00 + 2.00	298.58	23.50	37.31	86.19
−2.00 + 1.00	260.19	20.48	57.79	62.69
−1.00 + 0.5	186.05	14.64	72.43	42.21
−0.5 + 0.15	133.12	10.48	82.91	27.57
−0.15 + 0.106	77.36	6.09	89.00	17.09
−0.106 + 0.075	54.74	4.30	93.30	11.00
−0.075	85.02	6.70	100.00	6.70
Total	1270.59	100.00	—	—

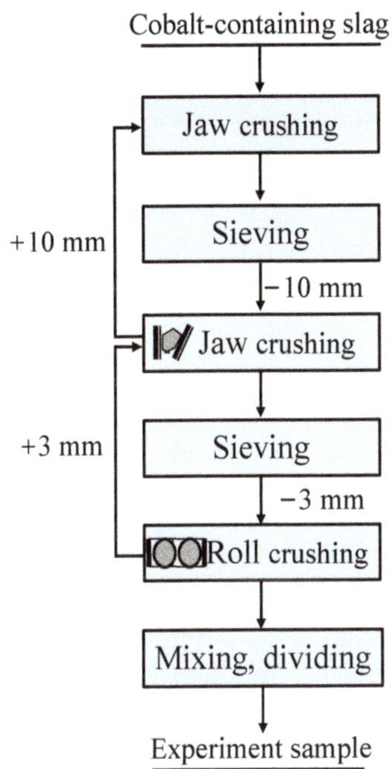

Figure 3. Pretreatment process of experimental sample of cobalt-containing slag.

2.2.2. Reduction Roasting Experiments

In each experiment, 50 g of cobalt-containing slag was mixed with a reducing agent, with anthracite used as the reductant. The amount of anthracite was varied between 30% and 50% of the weight of the cobalt-containing slag. The thoroughly mixed sample was placed in a crucible and then transferred to a muffle furnace. The furnace was heated to the experimental target temperature at a heating rate of 10 °C/min at room temperature, where carbothermal reduction roasting was carried out. After reduction roasting, the sample was taken out at the experimental target temperature and naturally cooled to room temperature in air. The roasting product was then crushed and ball-milled to achieve a particle size distribution where 75% passed −0.075 mm, while the remaining 25% ranged between −2 mm and −0.075 mm. Finally, the cobalt concentrate was obtained through magnetic separation, with the magnetic field intensity set to 95.54 kA/m. The experimental process of reduction roasting is shown in Figure 4.

Static reduction experiment: The 300 g of cobalt-containing slag was mixed with 90–150 g of anthracite. The mixture was placed in an 800 mL crucible and transferred to a muffle furnace. The furnace was heated to the target temperature at a rate of 10 °C/min to perform carbothermal reduction roasting. After roasting, the sample was removed and allowed to cool naturally. Rotary kiln dynamic experiment: The 500 g of cobalt-containing slag was mixed with 150–250 g of anthracite and placed in a laboratory rotary kiln with a diameter of 90 mm. The kiln was heated to the target temperature at a rate of 10 °C/min to carry out dynamic reduction roasting. After roasting, the sample was removed and allowed to cool naturally. The roasted samples obtained from both static and dynamic reduction roasting were subjected to ball-milling to achieve a particle size of −0.075 mm (accounting for 75% of the total particles). Magnetic separation was then performed at a magnetic field intensity of 95.54 kA/m to obtain cobalt concentrate.

The cobalt, iron, and copper contents in the cobalt-containing slag and the magnetic separation concentrate were measured to calculate the recovery rates of Co, Fe, and Cu. The measured concentrations of Co, Fe, and Cu in the cobalt concentrate represent the grades of these elements. The recovery rates of Co, Fe, and Cu were calculated using the following:

$$R = 1 - \frac{m_1 c_1}{m_0 c_0}$$

where m_0 and m_1 represent the mass of the cobalt-containing and residue slag, respectively; c_0 represents the Co, Fe, and Cu contents in the converter slag; c_1 represents the Co, Fe, and Cu contents in the residue slag.

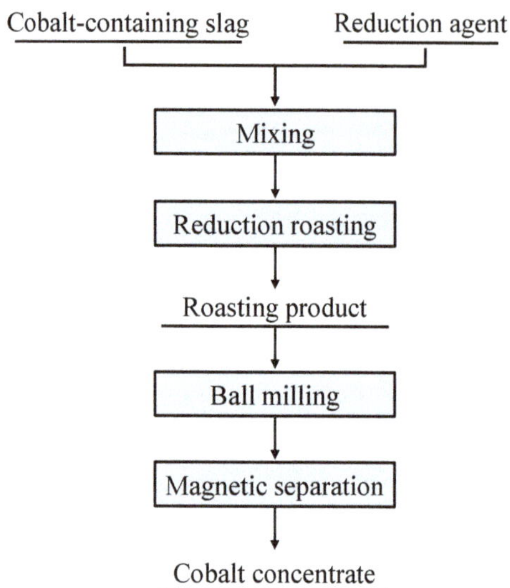

Figure 4. Reduction roasting experiment process.

2.2.3. Analytical Methods

The chemical component of the cobalt-containing slag samples was determined by inductively coupled plasma atomic emission spectrum (ICP-AES, IRIS Advantage Radial, PerkinElmer, Waltham, MA, USA) and X-ray fluorescence spectroscopy (XRF, Axios advanced, Thermo Fisher Scientific, Waltham, MA, USA). And the main phase compositions of the cobalt-containing slag were detected by an X-ray diffractometer (XRD, Simens D500 automatic X-ray diffractometer, Siemens AG, Berlin, Germany) with a copper target and operated at 40 kV and 250 mA in step mode with a 0.02° 2θ step and a count time of 0.5 s per step over a 2θ range from 10° to 80°. The morphology and the distribution of elements in cobalt-containing slag and roasting products were observed under a scanning electron microscope (SEM, JEOL, Tokyo, Japan) equipped with an energy-dispersive spectrometer (EDS, BRUKER, Bremen, Germany), operated at 20.0 kV after gold spraying of the samples. The cobalt-containing slag and roasting product samples were observed in the optical microscope (Leica DM4500P, Leica Camera AG, Solms, Germany) by using the reflected plane polarized light. According to the different colors of minerals and pores presented in the optical micrographs, the software of Image-Pro Plus 6.0 was used to quantitatively determine the mineral compositions via the area calculation method.

3. Results and Discussion

3.1. Thermodynamic Analysis

The composition of cobalt-containing slag is complex, and the reactions of Fe_3O_4, FeO, Fe_2SiO_4, $CoFe_2O_4$, CoO, and CuO with fixed carbon at atmospheric pressure are shown in Equations (1)–(6). The Gibbs free energy variations in these reactions with temperature were calculated using HSC Chemistry 6 software, and the results are presented in Figure 5a. Based on the Gibbs free energy values, the initial reduction temperatures of Fe_3O_4, FeO, and CoO by carbon were determined to be 675 °C, 722 °C, and 498 °C, respectively. Theoretically, CuO can be reduced to metallic Cu at room temperature. Furthermore, according to the equilibrium diagrams in Figure 5c,d, the primary attachment carriers of Co, such as iron silicates and cobalt–iron oxides, can undergo reduction reactions at 792 °C and 722 °C, respectively, reducing Co and Fe to their metallic forms.

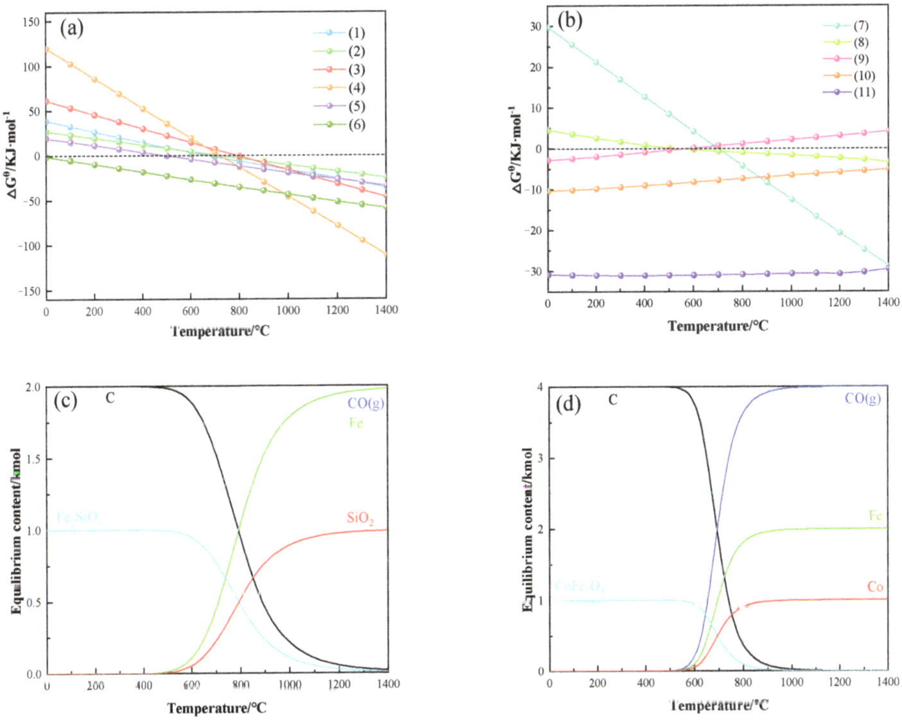

Figure 5. (**a**) Gibbs free energy of Equations (1)–(6) reduction at different temperatures; (**b**) Gibbs free energy of Equations (7)–(11) reduction at different temperatures; (**c**) theoretical carbon reduction equilibrium diagram of Fe_2SiO_4; (**d**) theoretical carbon reduction equilibrium diagram of $CoFe_2O_4$.

During the reduction process, carbon undergoes the Boudouard reaction, generating CO gas (Equation (7)), which subsequently acts as a reducing agent for metal oxides, as illustrated in Equations (8)–(11). As shown in Figure 5b, calculations indicate that the Boudouard reaction becomes thermodynamically favorable at temperatures above 700 °C. As the equilibrium partial pressure of CO_2 in the system increases, it promotes the gasification of carbon, shifting the reduction pathway of metal oxides from solid–solid reduction to gas–solid reduction, thereby enhancing the overall reduction efficiency.

$$2Fe_3O_4 + C = 6FeO + CO_2(g) \tag{1}$$

$$FeO + C = Fe + CO \quad (2)$$

$$Fe_2SiO_4 + 2C = 2Fe + SiO_2 + 2CO(g) \quad (3)$$

$$CoFe_2O_4 + 4C = Co + 2Fe + 4CO(g) \quad (4)$$

$$CoO + C = Co + CO(g) \quad (5)$$

$$CuO + C = Cu + CO(g) \quad (6)$$

$$C + CO_2(g) = 2CO(g) \quad (7)$$

$$Fe_3O_4 + CO(g) = 3FeO + CO_2(g) \quad (8)$$

$$3FeO + CO(g) = Fe + CO_2(g) \quad (9)$$

$$CoO + CO(g) = Co + CO_2(g) \quad (10)$$

$$CuO + C = Cu + CO_2(g) \quad (11)$$

3.2. *Effect of Reduction Temperature*

The selection of the reduction temperature is crucial. The reasonable selection of the reduction temperature is directly related to the metal grade and metal recovery rate of the magnetic separation concentrate. When the reduction agent ratio was fixed at 40%, the reduction time was 2 h, and the particle size of the cobalt-containing slag was −3 mm, the effect of the reduction temperature on the recovery rates of Co, Cu, and Fe and the grades of Co, Cu, and Fe was investigated. The result is shown in Figure 6.

It can be observed that with the increase in temperature, the grades of Co, Cu, and Fe showed an upward trend, reaching high points of 4.02%, 2.48%, and 81.33%, respectively, at 1150 °C. As the temperature rose, the grades of Co, Cu, and Fe all decreased. The recovery rates of the Co, Cu, and Fe metals increased with the increasing temperature, but the increase was smaller between 1150 °C and 1200 °C. The reduction in cobalt from its oxide form was nearly complete at between 1000 °C and 1150 °C, resulting in a sharp increase in the cobalt recovery rate. Therefore, considering both the metal grades and metal recovery rate indicators, the reduction temperature was selected as 1150 °C.

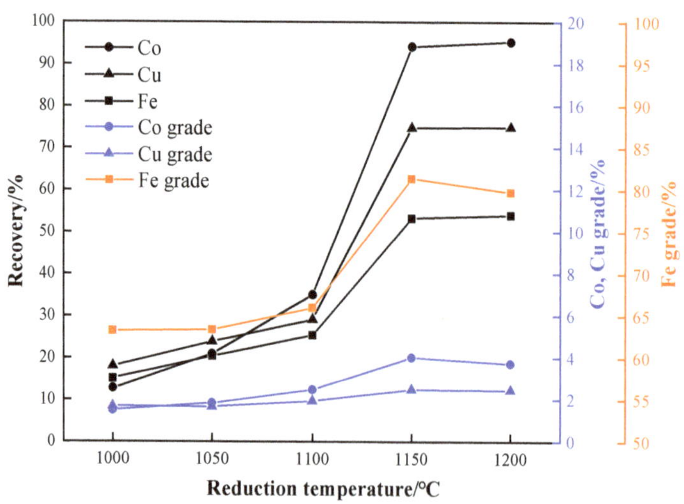

Figure 6. Effect of reduction temperature on metal recovery.

3.3. Effect of Reduction Agent Ratio

The appropriate reduction agent ratio can regulate the reduction degree of Co and Fe to increase the enrichment ratio of cobalt concentrate. The effects of the reducing agent ratio on the recoveries of Co, Cu, and Fe and the grades of Co, Cu, and Fe were investigated when the reduction temperature was 1150 °C, the reduction time was 2 h, and the particle size of the cobalt-containing slag was −3 mm.

It can be seen from Figure 7 that with the increase in the reduction agent ratio, the Co grade of concentrate has a slight change of about 4%, which shows a slight downward trend. The Co grade gradually decreased as the reducing agent ratio increased to 40% and then remained unchanged. The Fe grade reached 81.33% when the reducing agent ratio was 40%. The recovery rates of Co, Cu, and Fe first gradually increased and then decreased with the increase in the amount of reduction agent added. When the reduction agent ratio was 40%, the highest recovery rates of Co, Cu, and Fe were 94.17%, 74.8%, and 53.27%, respectively. When the reduction agent ratio exceeded 40%, the excessive reduction of iron oxides led to the formation of non-magnetic phases, which are hard to separate during the subsequent magnetic separation process, resulting in a decrease in the iron grade. Once cobalt oxide is reduced to metallic cobalt, its stability is high, making it resistant to further reduction into other lower-valent compounds, thus maintaining a stable cobalt grade. Therefore, the reduction agent ratio was determined to be 40%.

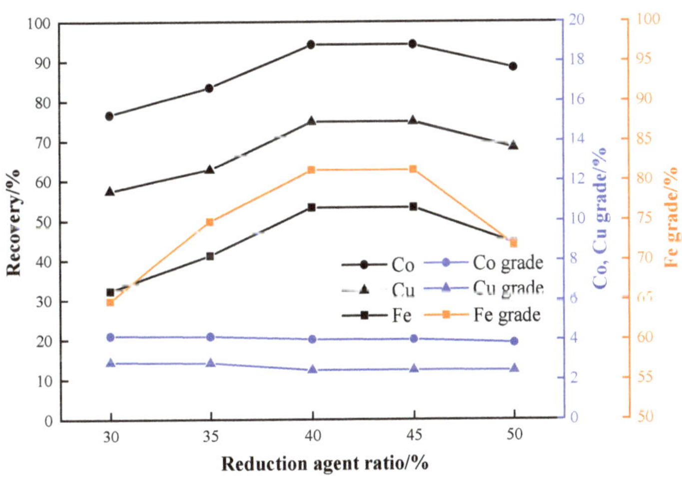

Figure 7. Effect of the reduction agent ratio on metal recovery.

3.4. Effects of Reduction Time

Metal oxides begin to be reduced under specific temperature conditions, and the extent of reduction is closely related to the reduction time. When the reduction temperature is fixed at 1150 °C, the reduction agent ratio is 40%, and the particle size of the cobalt-containing slag is −3 mm, the effect of the reduction time on the recovery rates of Co, Cu, and Fe and the grades of Co, Cu, and Fe is investigated.

It can be seen from the analysis of Figure 8 that the grades of Co, Cu, and Fe showed an upward trend before 2 h, reached the highest of 4.02%, 2.48%, and 81.33%, respectively, at 2 h, then showed an inflection point at 2 h, and then declined. The recovery rate of Cu and Fe was increasing. The recovery rate of Co first increased with time and then decreased slightly after 2 h. Therefore, the reduction time was determined to be 2 h.

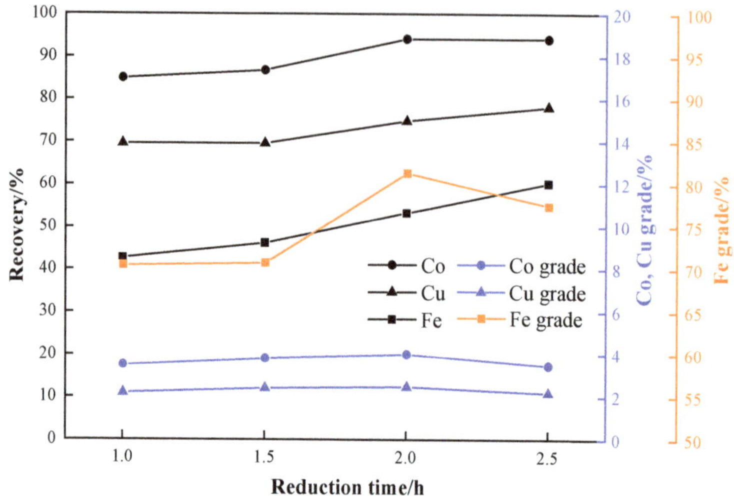

Figure 8. Effect of reduction time on metal recovery.

3.5. Effects of Particle Size

As can be seen from the data in Figure 9, there is no significant difference in the reduction roasting effect when the material particle size is −3 mm, −6 mm, −10 mm, or −13 mm. The Co grade of the concentrate and the Co metal recovery rate are basically the same. The Co grade is basically around 4%, and the Co recovery rate is around 95%. Smaller particle sizes generally improve recovery efficiency by providing a larger reaction surface area and enhancing the diffusion capacity of the reducing agent. However, within the particle size range of −3 mm to −13 mm, no significant difference in cobalt recovery was observed. This suggests that within this range, the particle size of the cobalt-containing slag has little impact on the efficiency of the reduction process. The Cu and Fe grades also do not change much. The Cu and Fe recovery rates first increase and then slightly decrease as the particle size of the cobalt-containing slag decreases. Considering the metal grade and recovery rate, in order to maintain the consistency of the experiment, the −3.0 mm material particle size was selected.

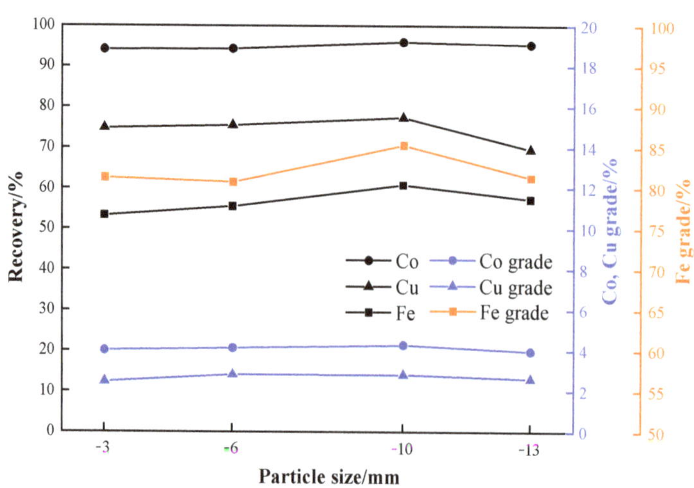

Figure 9. Effect of particle size on metal recovery.

3.6. Reduction Verification Experiment

Under the optimal conditions of the reduction temperature of 1150 °C, the reduction agent ratio of 40%, the reduction time of 2 h, and the particle size of −3 mm, the static reduction experiment of three samples of cobalt-containing slag was carried out in the muffle furnace, and the experiment results are shown in Figure 10. As can be seen from Figure 10, the cobalt grade is between 3.93 and 4.30%, and the recovery rate is between 90.49% and 92.66% in the three static reduction experiments under the best conditions. At the same time, the rotary kiln dynamic reduction experiment was carried out under the same optimal reduction conditions. It can be seen in Figure 11 that the cobalt grade and cobalt recovery rate reached 4.26–4.41% and 90.67–93.67%, respectively. The main reason for the better results of the dynamic reduction experiment is that the mass and heat transfer of materials in the high-temperature reduction environment during the rotation of the rotary kiln are higher than in the muffle furnace. In summary, the static and dynamic experiment results are stable and can provide technical support for industrialization.

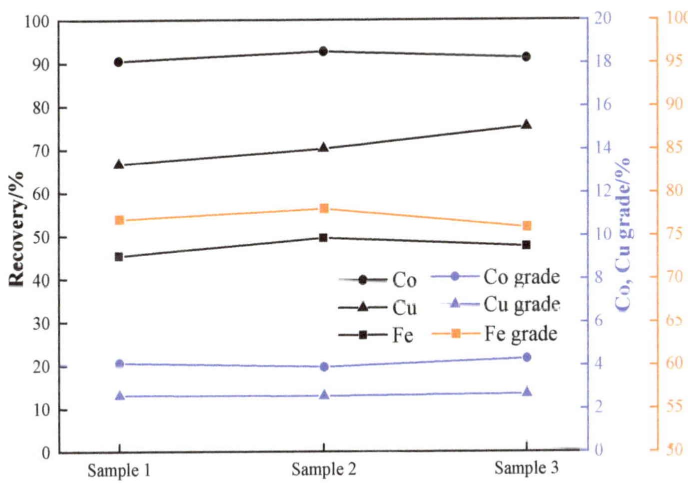

Figure 10. Static reduction experiment under optimal conditions.

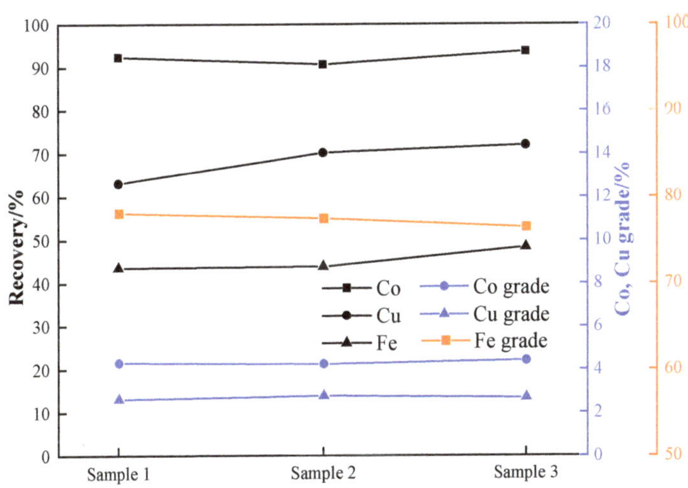

Figure 11. Rotary kiln dynamic experiment under optimal conditions.

3.7. Migration of Cobalt

The cobalt-containing slag and the roasting product were detected by optical microscopy, and the mineral content was statistically analyzed. Then, the cobalt distribution rate was obtained by combining the statistical results of mineral content and scanning electron microscopy spectroscopy, and the results are shown in Table 5 and Figure 12. It can be seen that the main minerals of cobalt-containing slag are iron-containing silicate, vitreous minerals, magnetite, and copper sulfide minerals, followed by a small amount of cobalt–nickel sulfide, copper metal, and pyrrhotite. The iron-containing silicate and vitreous minerals in cobalt-containing slag are cemented to form a matrix, and copper sulfide minerals and magnetite are embedded in the matrix in granular forms of varying thicknesses. Cobalt–nickel sulfide and pyrrhotite are mostly wrapped in fine particles in copper sulfide minerals, and copper metal is distributed in copper sulfide minerals and the matrix.

As shown in the table, the cobalt in the cobalt-containing slag is mainly attached to the iron-containing silicate, and the distribution rate of cobalt is as high as 96.93%. The cobalt–iron alloy accounted for 18.34% of the total mineral content in the roasting products, the iron-containing silicate content was 51.13%, but the cobalt distribution rate was low. Cobalt was mainly distributed in cobalt–iron alloy after reduction roasting, accounting for 82.10%, which indicates that cobalt oxide is reduced and dissociated from iron-containing silicate at a high temperature during the reduction roasting process. Cobalt forms a strong magnetic cobalt–iron alloy with iron, which can be recovered by magnetic enrichment.

Table 5. Distribution of cobalt in cobalt-containing slag and roasting product (%).

Mineral	Content of Cobalt-Containing Slag	Cobalt Distribution Rate	Content of Roasting Product	Cobalt Distribution Rate
Cobalt–nickel sulfide	0.10	2.52	-	-
Copper sulfide	1.87	0	0.75	0.45
Magnetite	5.01	0.49	0.83	0
Pyrrhotite	0.02	0.06	-	-
Metallic copper	0.05	0	0.08	0.02
Iron-containing silicate	77.88	96.93	51.13	15.79
Vitreous mineral	15.07	0	28.87	1.64
Cobalt–iron alloy	-	-	18.34	82.10

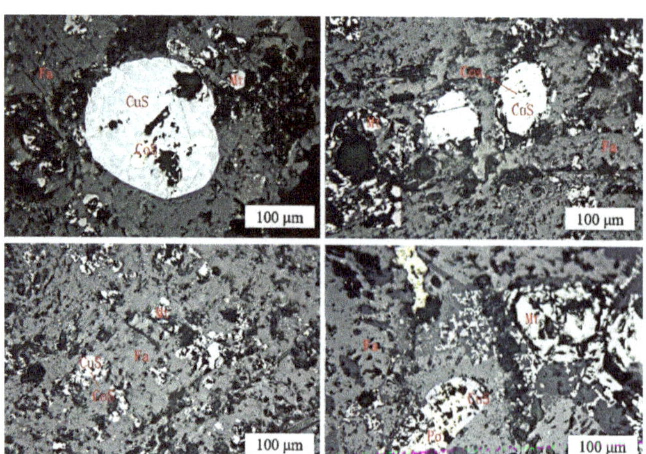

Figure 12. Microstructure of cobalt-containing slag. Mt—magnetite; Po—pyrrhotite; Fa—iron-containing silicate.

4. Conclusions

(1) The primary minerals in cobalt-containing slag include iron silicate minerals, glassy minerals, magnetite, and copper sulfide minerals, with smaller amounts of cobalt-nickel sulfide minerals, metallic copper, and pyrrhotite. The main cobalt-bearing minerals are cobalt–nickel sulfide minerals, pyrrhotite, iron silicate minerals, and magnetite. Cobalt is predominantly hosted in the iron silicate minerals.

(2) The optimal process conditions for the carbothermic reduction of cobalt-containing slag, based on experimental results, are as follows: a reduction temperature of 1150 °C, a reductant ratio of 40%, a reduction time of 2 h, a particle size of −3 mm, a grinding fineness of −0.075 mm with an 80% passing rate, and a magnetic field strength of 95.54 kA/m. The results show that the grades of cobalt, copper, and iron in the concentrate after magnetic separation are 4.02%, 2.48%, and 81.33%, respectively, and the recoveries are 94.17%, 74.80%, and 53.27%.

(3) The static reduction in a muffle furnace and the dynamic reduction in a rotary kiln were validated under the optimal conditions, resulting in stable cobalt grade and recovery rates, which are suitable for industrial applications. The cobalt grade is between 3.93 and 4.30%, and the recovery rate is between 90.49% and 92.66% according to the static reduction experiment. The cobalt grade and cobalt recovery rate reached 4.26–4.41% and 90.67–93.67%, respectively, with the rotary kiln dynamic reduction experiment.

Author Contributions: Conceptualization, J.G., J.P., J.Z., Q.Z., G.H., Y.L. and H.Y.; methodology, J.G., J.P., J.Z., Q.Z. and G.H.; software, J.G., J.P. and Q.Z.; validation, J.G., J.P., J.Z. and G.H.; formal analysis, J.G., J.P., J.Z., G.H. and H.Y.; investigation, J.G., Q.Z., Y.L. and H.Y.; resources, J.G., J.P., J.Z., Q.Z., G.H., Y.L. and H.Y.; data curation, J.G., J.P., J.Z., Q.Z., G.H., Y.L. and H.Y.; writing—original draft preparation, J.G., J.P., J.Z., Q.Z., G.H., Y.L. and H.Y.; writing—review and editing, J.G., J.P. and Q.Z.; visualization, J.G., J.P., J.Z., Q.Z., G.H., Y.L. and H.Y.; supervision, J.G., J.P. and J.Z.; project administration, J.G. and J.Z. All authors have read and agreed to the published version of the manuscript.

Funding: This research received no external funding.

Data Availability Statement: The original contributions presented in the study are included in the article, further inquiries can be directed to the corresponding authors.

Conflicts of Interest: Authors Jiachen Gong, Jingfu Zhao, Guansheng Hao, Yan Liu and Helei Yu were employed by the company CNMC Shenyang Research Institute of Nonferrous Metals Co., Ltd. The remaining authors declare that the research was conducted in the absence of any commercial or financial relationships that could be construed as a potential conflict of interest.

References

1. Song, Y.; Zhang, Z.; Cheng, J.; Zeng, A.; Zhang, Y. Effects of technical advance on cobalt supply and its supply trends: A perspective of the whole industrial chains. *Resour. Conserv. Recycl.* **2024**, *209*, 107760. [CrossRef]
2. Chandra, M.; Yu, D.; Tian, Q.; Guo, X. Recovery of Cobalt from Secondary Resources: A Comprehensive Review. *Miner. Process. Extr. Metall. Rev.* **2021**, *43*, 679–700. [CrossRef]
3. Santoro, L.; Tshipeng, S.; Pirard, E.; Bouzahzah, H.; Kaniki, A.; Herrington, R. Review of Economic, Technical and Environmental Aspects of Electric Vehicles. *Miner. Eng.* **2019**, *137*, 277–289. [CrossRef]
4. Ettler, V.; Mihaljevič, M.; Drahota, P.; Kříbek, B.; Nyambe, I.; Vaněk, A.; Penížek, V.; Sracek, O.; Natherová, V. Cobalt-bearing copper slags from Luanshya (Zambian Copperbelt): Mineralogy, geochemistry, and potential recovery of critical metals. *J. Geochem. Explor.* **2022**, *237*, 106987. [CrossRef]
5. Zhang, Y.; Fan, Y.; Liu, Y.; Zhou, T.; Ou, B. Distribution and enrichment processes of cobalt in the Longqiao iron skarn deposit in Eastern China. *Ore Geol. Rev.* **2024**, *174*, 106277. [CrossRef]
6. Horn, S.; Gunn, A.G.; Petavratzi, E.; Shaw, R.A.; Eilu, P.; Törmänen, T.; Bjerkgård, T.; Sandstad, J.S.; Jonsson, E.; Kountourelis, S.; et al. Cobalt resources in Europe and the potential for new discoveries. *Ore Geol. Rev.* **2021**, *130*, 103915. [CrossRef]

7. Santoro, L.; Tshipeng, S.; Pirard, E.; Bouzahzah, H.; Kaniki, A.; Herrington, R. Mineralogical reconciliation of cobalt recovery from the acid leaching of oxide ores from five deposits in Katanga (DRC). *Miner. Eng.* **2019**, *137*, 277–289. [CrossRef]
8. Dehaine, Q.; Tijsseling, L.T.; Glass, H.J.; Törmänen, T.; Butcher, A.R. Geometallurgy of cobalt ores: A review. *Miner. Eng.* **2021**, *160*, 106656. [CrossRef]
9. Shannak, S.D.; Cochrane, L.; Bobarykina, D. Strategic analysis of metal dependency in the transition to low-carbon energy: A critical examination of nickel, cobalt, lithium, graphite, and copper scarcity using IEA future scenarios. *Energy Res. Soc. Sci.* **2024**, *118*, 013773. [CrossRef]
10. Savinova, E.; Evans, C.; Lèbre, É.; Stringer, M.; Azadi, M.; Valenta, R.K. Will global cobalt supply meet demand? The geological, mineral processing, production and geographic risk profile of cobalt. *Resour. Conserv. Recycl.* **2023**, *190*, 106855. [CrossRef]
11. Wesselkämper, J.; Dahrendorf, L.; Mauler, L.; Lux, S.; von Delft, S. Towards circular battery supply chains: Strategies to reduce material demand and the impact on mining and recycling. *Resour. Policy* **2024**, *95*, 105160. [CrossRef]
12. Tisserant, A.; Pauliuk, S. Matching global cobalt demand under different scenarios for co-production and mining attractiveness. *J. Econ. Struct.* **2016**, *5*, 4. [CrossRef]
13. Liu, Y.; Xia, Q.; Duan, J.; Dai, J.; Wu, S.; Zhao, Z. Geochemical anomalies of critical metals in the Eastern Kunlun Orogenic Belt, China: Implications for nickel and cobalt mineral exploration. *Ore Geol. Rev.* **2024**, *171*, 106168. [CrossRef]
14. Seck, G.S.; Hache, E.; Barnet, C. Potential bottleneck in the energy transition: The case of cobalt in an accelerating electro-mobility world. *Resour. Policy* **2022**, *75*, 102516. [CrossRef]
15. Piçarra, A.; Annesley, I.R.; Otsuki, A.; de Waard, R. Market assessment of cobalt: Identification and evaluation of supply risk patterns. *Resour. Policy* **2021**, *73*, 102206. [CrossRef]
16. Chen, Z.; Zhang, L.; Xu, Z. Tracking and quantifying the cobalt flows in mainland China during 1994–2016: Insights into use, trade and prospective demand. *Sci. Total Environ.* **2019**, *672*, 752–762. [CrossRef]
17. Yu, Z.; Wang, Y.; Ma, X.; Shuai, C.; Zhao, Y. How critical mineral supply security affects China NEVs industry? Based on a prediction for chromium and cobalt in 2030. *Resour. Policy* **2023**, *85*, 103861. [CrossRef]
18. Liu, W.; Li, X.; Liu, C.; Wang, M.; Liu, L. Resilience assessment of the cobalt supply chain in China under the impact of electric vehicles and geopolitical supply risks. *Resour. Policy* **2023**, *80*, 103183. [CrossRef]
19. Wang, Y.; Zhang, S.A.; Wang, L.; Qin, W.; Han, J. An unconventional approach for the efficient recovery of iron, cobalt, copper and silicon from copper slag. *J. Hazard. Mater.* **2024**, *476*, 135168. [CrossRef]
20. Zhang, Z.; Zhang, Y.; Ning, Y.; Wei, G.; Qu, J.; Yu, Z.; Liu, X. Process and mechanism of Zn and Co leaching from cobalt xanthate slag enhanced by oxidative roasting. *Process Saf. Environ. Prot.* **2024**, *192*, 1249–1259. [CrossRef]
21. Godirilwe, L.L.; Haga, K.; Altansukh, B.; Jeon, S.; Danha, G.; Shibayama, A. Establishment of a Hydrometallurgical Scheme for the Recovery of Copper, Nickel, and Cobalt from Smelter Slag and Its Economic Evaluation. *Sustainability* **2023**, *15*, 10496. [CrossRef]
22. Li, L.; Xiao, Y.; Lei, Y.; Xu, J.; Xu, Z. An approach of cobalt recovery from waste copper converter slags using pig iron as capturing agent and simultaneous recovery of copper and tin. *Waste Manag.* **2023**, *165*, 1–11. [CrossRef]
23. Li, H.; Guo, X.-Y.; Shi, Y.; Wang, C.; Yu, D.-W.; Tian, Q.-H. Selective hydrogen reduction of binary iron-cobalt chlorides. *J. Cent. South Univ.* **2024**, *30*, 3991–4003. [CrossRef]
24. Li, Y.; Chang, C.; Jie, Y.; Jin, W.; Chen, Y.; Wan, X.; Tang, C.; Yang, S. Thermodynamic Phase Conversion Mechanism on Copper–Cobalt Slag Cleaning Process Using Gypsum Wastes as Sulfurizing Agent. *J. Sustain. Metall.* **2021**, *7*, 1643–1653. [CrossRef]
25. Huang, Y.; Duan, Z.; Bai, N.; Wang, H.; Cao, Y.; Song, X.; Peng, W.; Zhu, X. Highly selective dissolution and synchronous extraction of zinc from zinc-cobalt slag by an ionic liquid [Hbet][Tf_2N]–H_2O system: A novel method for separating zinc and cobalt. *J. Clean. Prod.* **2021**, *315*, 128301. [CrossRef]
26. Zhao, H.; Ma, B.; Hong, S.; Huang, H.; Liu, F.; Sohn, H.Y. Recovery of Copper and Cobalt from Converter Slags via Reduction–Sulfurization Smelting Using Spent Pot Lining as the Reductant. *ACS Sustain. Chem. Eng.* **2021**, *9*, 4234–4246. [CrossRef]
27. Alvial-Hein, G.; Mahandra, H.; Ghahreman, A. Separation and recovery of cobalt and nickel from end of life products via solvent extraction technique: A review. *J. Clean. Prod.* **2021**, *297*, 126592. [CrossRef]
28. Yang, X.; Zhang, J.; Zhang, J.; Hu, J.; Li, J.; Zhang, L.; Chen, Y.; Wang, C. Efficient Recovery of Copper and Cobalt from the Matte–Slag Mixture of ISA Furnace by Injection of Coke and Pyrite. *Metall. Mater. Trans. B* **2018**, *49*, 3118–3126. [CrossRef]
29. Phiri, T.C.; Singh, P.; Nikoloski, A.N. Mineralogical Characterisation of Copper Slag and Phase Transformation after Carbocatalytic Reduction for Hydrometallurgical Extraction of Copper and Cobalt. *Metals* **2024**, *14*, 1119. [CrossRef]
30. Rudnik, E.; Burzyńska, L.; Gumowska, W. Hydrometallurgical recovery of copper and cobalt from reduction-roasted copper converter slag. *Miner. Eng.* **2009**, *22*, 88–95. [CrossRef]
31. Zhai, X.-J.; Li, N.-J.; Zhang, X.; Fu, Y.; Jiang, L. Recovery of cobalt from converter slag of Chambishi Copper Smelter using reduction smelting process. *Trans. Nonferrous Met. Soc. China* **2011**, *21*, 2117–2121. [CrossRef]

32. Bhatti, S.A.; Qiao, X.-C. Synergistic effect of carbothermal reduction and sodium salts leaching in the process of iron recovery from copper slag. *Process Saf. Environ. Prot.* **2025**, *193*, 170–182. [CrossRef]
33. *GB/T 212-2008*; Industrial Analysis of Coal. National Standard of the People's Republic of China: Beijing, China, 2021.

metals

MDPI

Article

Extraction of REEs and Other Elements from Estonian Graptolite-Argillite and Phosphorite Mineral Acid Solutions

Silvester Jürjo, Liis Siinor, Carolin Siimenson, Ove Oll and Enn Lust *

Institute of Chemistry, University of Tartu, 50411 Tartu, Estonia; carolin.siimenson@ut.ee (C.S.)
* Correspondence: enn.lust@ut.ee

Abstract: The chemical extraction of rare-earth elements (REEs) from Estonian graptolite-argillite (GA) and phosphate rock (Phosphorite, PH) samples has been conducted and analyzed. For the initial leaching process, HCl and HNO_3 with different concentrations were used to extract REEs from GA and PH. Different extraction agents, including ionic liquids, were examined for the extraction of REEs from acidic aqueous solutions in the liquid–liquid extraction step. After leaching and extraction, all samples were characterized using the inductively coupled plasma mass spectrometry method (ICP-MS/MS). The highest REE extraction efficiencies from GA were established with 1-ethyl-3-methyl imidazolium diethyl phosphate ($EMImDEPO_4$) and from PH using bis(2-ethylhexyl) phosphate (D2EHPA).

Keywords: Estonian graptolite-argillite ore; phosphorite ore; ionic liquids; liquid–liquid extraction; extraction of REEs

1. Introduction

Rare-earth elements (REEs) include 15 lanthanide elements (Z = 57 to 71) together with scandium (Z = 21) and yttrium (Z = 39), as stated in the International Union of Pure and Applied Chemistry (IUPAC) [1–4]. Due to their high reactivity and chemical similarity, REEs are complicated to separate and refine into pure metal. As a result, efforts to develop efficient separation processes began early in the 20th century [5].

Nowadays, REEs have an enormous impact on societies' daily lives since it is impossible to develop any high-tech technology that would not contain any REEs [1,6–8]. Energy storage devices have become indispensable for storing and maintaining a continuous power supply to meet established renewable energy targets. Furthermore, REE-based permanent magnets are extensively used in wind turbines and electric vehicles [9–11]. Current supply chains are mainly controlled and fulfilled by the Chinese REE industry [12,13]. China dominates the REE industry by producing more than 90% of rare-earth requirements. Therefore, recycling REEs is currently necessary to alleviate market fluctuations [14].

Black shale formations, known as graptolite-argillite, are found along the northern coast of Estonia and on Vormsi and Hiiumaa islands [15,16]. The thickness of GA reaches up to 7.4 m in Northwest Estonia and decreases towards the east and south [15,16]. Black shales can be divided into classes based on the degree of commodity metal accumulation: black shale (0.01%), metalliferous black shales (0.1%), and hyper-enriched black shales (1%) [17]. The distribution of metals in GA has a complex pattern, but Mo, V, and U are the most characteristic. Moreover, GA contains high variability in the trace metal composition, including heterogeneous REE patterns [17–19].

Presumably, the largest unexploited sedimentary phosphate rock reserves in the European Union are also located in Estonia (about 700 million tons of P_2O_5) [20]. Estonian phosphate rock contains low amounts of Cd (1–5 ppm) and depleted U content compared to other worldwide deposit locations, but the total content of REEs has been identified with an average of 1500–2000 ppm [20].

REE leaching used in primary REE production is integral to hydrometallurgical REE processing [21]. These processes are defined as noncatalytic heterogeneous reactions between liquid and solid phases, ranging from acid leaching (H_2SO_4, HCl, or HNO_3) of primary ores to leaching with NaCl or $(NH4)_2SO_4$ of ion-adsorbed clays [22,23]. For industrial design and optimization processes, the kinetics aspect of the leaching is considered to be highly relevant, including various types of diffusion and chemical reaction mechanisms [24]. Therefore, establishing the most optimal pathway would significantly enhance the leaching process's effectiveness and consider the environmental aspects.

Extraction of cations is an efficient technique for separating rare earths from acid solutions—it is relatively simple, rapid, and applicable in various concentrations, pH, and temperature ranges [25,26]. For example, triphenyl arsine in kerosene can extract yttrium and ytterbium from sulfuric acid medium. Ionic liquids (ILs) are studied as extraction agents, emphasizing the existing technologies and the principles of sustainable and green chemistry [27,28]. Common acidic extractants (organic phosphonic acids and carboxylic acids) require saponification, resulting in large amounts of wastewater [29]. Thus, designing alternative extractants is a matter of great urgency. The non-functional ILs showed better extraction ability and selectivity than conventional organic solvent systems [30].

This paper has two main novel objectives regarding the high amount of black shale formations and large phosphate rock reserves in Estonia. Firstly, develop a leaching process for Estonian graptolite-argillite and phosphorite ore samples. Secondly, to extract REEs (separate HREEs and LREEs) and other distributed rare metal cations (DRM) from prepared aqueous solutions using ionic liquids and other extraction agents. The concentration of REEs and DRM was determined before and after extraction by ICP-MS/MS to calculate the respective extraction efficiencies. The influence of different molar concentrations of acids has been discussed.

2. Experimental Section

2.1. Leaching Step of GA and PH Samples

The raw GA powder samples from Northeast Estonia (samples have been designated as Sillamäe, Sõtke top, Sõtke bottom, and Argillite from the collection) were homogenized using ball milling (stabilized ZrO_2 balls) and then treated with aqueous acidic solutions GA was leached using Aqua Regia, 1 M HCl, 1 M HNO_3, 3 M HCl, 3 M HNO_3, and 7.5 M HNO_3 aqueous solutions in 1:10 solid-to-liquid mass ratios in separate vials. Demineralized water was used to prepare the acid solutions. The mixtures of GA and acid were stirred (300 rpm) at 70 °C for 3 h, followed by a cooling step (to room temperature), centrifugation for 30 min, and decantation (solid residual separation).

PH leaching experiments contained phosphorite ore samples from Northeast Estonia (designated as Ülgase, Toolse, and Iru outcrop) that were processed in 100 mL 7.5 M HNO_3, 7.5 M HCl, 3 M HNO_3, and 3 M HCl acidic aqueous solutions for 24 h at room temperature (22 ± 1 °C) in 1:10 solid-to-liquid mass ratios. A relatively high acid concentration and leaching time were necessary to dissolve REEs from the PH ore completely. Thereafter, the solution was filtered to remove any undissolved substances.

Estonian concentrated phosphorite ore (Ülgase, Toolse, and Iru from Northern Estonia) and graptolite-argillite (Sillamäe, Sõtke top, Sõtke bottom, and Argillite from the collection) samples were obtained from geological collections at the Institute of Ecology and Earth Sciences, University of Tartu.

Chemicals: HNO_3 and HCl (ACS grade) were purchased from Merck (Burlington, MA, USA).

2.2. Details of the Extraction Process

Acidic aqueous feed solution from the leaching step and extractant were added to the separation funnel at a fixed 9:1 (aqueous/organic) ratio in all cases (GA and PH samples) at room temperature. The mixture was shaken carefully in a separation funnel for 1 min and kept still to reach the reaction equilibration of the extraction for one hour. The procedure was repeated two more times for the best separation efficiencies.

Extractants: Tributyl phosphate (TBP, ≥99.0%) from Merck; 1-ethyl-3-methyl imidazolium diethyl phosphate ($EMImDEPO_4$, ≥97%) and tricaprylylmethylammonium chloride (Aliquat 336, ≥97.0%) from Sigma-Aldrich; bis(2-ethylhexyl) phosphate (D2EHPA, ≥95.0%) from Merck.

2.3. Experimental Methods for Determining the Detailed Chemical Composition of GA and PH

Trace element analysis of samples after the leaching step and extraction process was performed using Agilent 8800 QQQ ICP-MS/MS (Agilent Technologies, Santa Clara, CA, USA). All untreated extraction solutions were diluted to a final dilution factor of 4000 before the analysis using a 2% HNO_3 solution. ICP-MS/MS measurements were performed in NoGas mode, and collision mode was used for masses that exhibited polyatomic spectral overlaps. Calibration standard solutions were used.

3. Results and Discussion

3.1. GA Raw Deposit Analysis After Leaching with Modified Aqua Regia Using ICP-MS/MS Method

The GA is an argillaceous rock enriched with organic matter that contains high concentrations of several elements (U, V, Mo, Pb, Zn, Cd, Cu, As) and very low concentrations of REEs [31]. The current study of geochemical generalizations of Estonian GA shows that the content of elements in the solution obtained from leaching is heterogeneous and highly dependent on the location and origin (Figure 1). The concentration of metal cations in GA obtained after the leaching process with modified Aqua Regia is in good agreement with the data by Soesoo et al. [15,32,33].

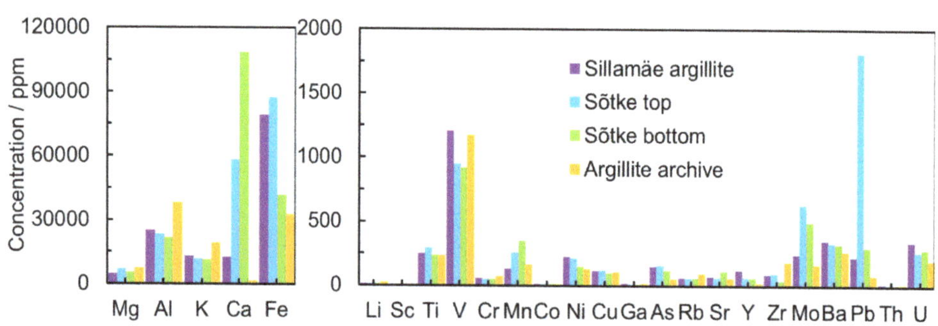

Figure 1. The concentration of elements after the leaching process in modified Aqua Regia aqueous solution from leaching for different GA deposits collected from Northern Estonia: Sõtke top, Sõtke bottom, Sillamäe, and Argillite.

The concentrations of U (198–280 ppm) and Th (10–13 ppm) were moderate in all GA samples. The REEs' total concentration (in the solution after the leaching process) in the Sillamäe sample reached 600 ppm but stayed between 200 and 350 ppm in other samples (Figure 2).

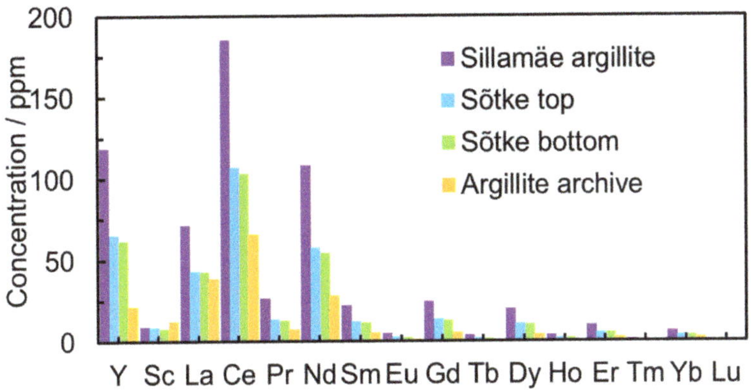

Figure 2. The concentration of REEs after the leaching process in modified Aqua Regia aqueous solution from leaching for different GA deposits from Northern Estonia, including Sõtke top, Sõtke bottom, Sillamäe, and Argillite.

3.2. GA and PH Composition Analysis After Leaching with HCl and HNO_3 Aqueous Solutions Using the ICP-MS/MS Method

For metal extraction, various forms of hydrometallurgical routes are used for black shale processing [17]. The leaching process with typical mineral acids (H_2SO_4, HCl) with higher concentrations increases the leachability of metals at atmospheric pressure [34]. In this study, H_2SO_4 was excluded due to earlier findings that most of the REEs transfer to the gypsum ($CaSO_4 \cdot 2H_2O$) during the sulfuric acid leaching [35]. Therefore, to analyze the influence of the acid concentration on the extraction efficiency (E %), HCl and HNO_3 with different concentrations were used for the GA and PH treatment processes.

The leaching process results of GA using acids with different concentrations show that the concentration of REEs decreases with the dilution of acid from 3 M to 1 M (with increasing pH of acid solution) based on the ICP-MS/MS measurement results (Figure 3). The lower leaching efficiency of REEs in more dilute acid solution can be explained by the lack of protons.

In the PH leaching experiments, the phosphorite ore samples from the Ülgase outcrop were leached in 7.5 M HNO_3, and the leaching efficiencies for different deposits are presented in Figure 4. High concentrations of La and Ce might be due to mineral grains of monazite (a mineral that contains rare-earth elements) in the sample [36].

A more systematic ICP-MS/MS data analysis (Figures 3 and 4) indicates that HNO_3 gives more concentrated REE solutions in both ore samples (GA and PH) since (NO_3^-) is a slightly harder Pearson base than Cl^-. Kinetically, hard acids are electrophiles that react more quickly with hard bases (nucleophiles), while soft acids react more rapidly with soft bases. This principle also allows for the prediction of displacement reaction kinetics. Thus, the softness of ions can be used to rationalize the relative stability of the cation–anion interaction in water, which can influence the stability of ion pairs and metal–ligand complexes in water relevant to hydrometallurgical processes [37].

Previous studies have also demonstrated that nitric acid is considered a mineral structure weakener of black shales [38]. Zhu et al. demonstrated that the Mo concentration can be increased with a higher HCl concentration in shales [39]. At the same time, the presence of HNO_3 resulted in 74% of the total shale Mo being leached and caused a preferential release of heavy Mo isotopes.

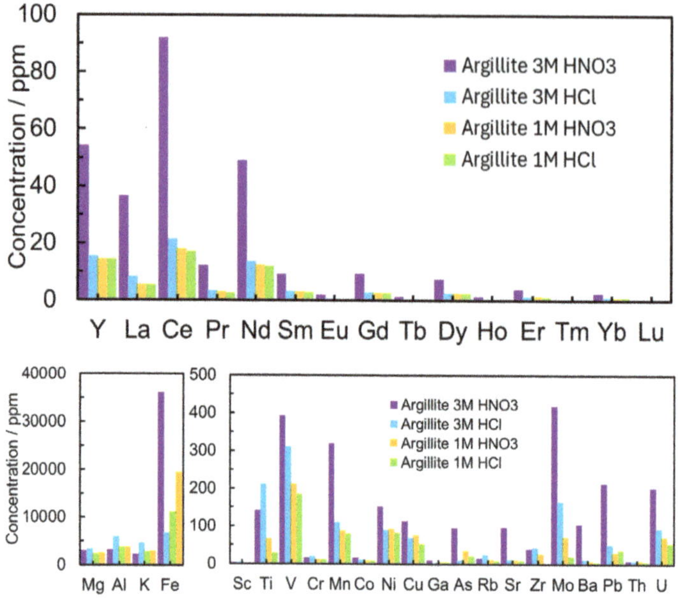

Figure 3. The concentration of elements in 1 M HCl, 3 M HCl, 1 M HNO_3, and 3 M HNO_3 aqueous solutions for different GA deposits from Argillite from the collection of the Institute of Geology UT.

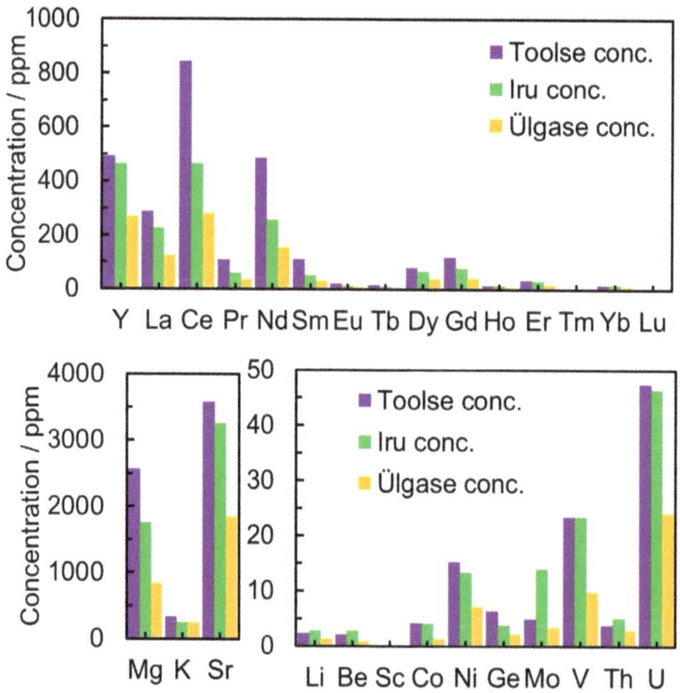

Figure 4. The concentration of elements in 7.5 M HNO_3 aqueous solution for different deposits collected from Toolse, Iru, and Ülgase outcrops.

3.3. Extraction Process of GA and PH Using Different Extractants

Based on the low waste materials production and natural resources conception accepted by the EU (recycling industry conditions), all REEs occurring in Estonian GA and PH should be extracted and separated [40].

All extraction agents applied were carefully chosen based on the literature data and our previous results [41–43]. Quaternary ammonium-based ionic liquid, Aliquat 336, modified with KNO_3 solution, has shown good extraction efficiency for La and Ce. The extraction efficiency with Aliquat 336 (NO_3) increased with an increase in the solution pH [44,45]. Therefore, the Aliquat 336(NO_3) extraction efficiency was also tested in the Estonian GA and PH systems. Phosphate-based extractants were applied to treat GA solutions, including TBP, EMImDEPO4, and Aliquat 336 (NO_3). Meanwhile, D2EHPA was used to extract REEs from PH aqueous acidic solutions in different concentrations, as it has been used in the literature before [46]. It is important to note that all REE concentrations (ppm) in this work depend on the sample's geological and mineralogical properties (mining pit) in addition to the experimental conditions of the leaching and extraction processes.

Extraction efficiency values were calculated using the previously introduced equation [43]:

$$E = \frac{C(initial) - C(final)}{C(initial)} \times 100\%$$

Table 1 demonstrates the extraction efficiencies of REEs from GA aqueous solutions. The REE extraction efficiencies of a common extractant, such as TBP, in Estonian GA systems are nonexistent compared to the 3 M HNO_3 solution before the extraction process. The calculated extraction efficiency values for ionic liquid, $EMImDEPO_4$, are excellent, higher than 98% for all REEs. It suggests that $DEPO_4^-$ strongly interacts with REEs in the acidic aqueous solution of GA. Based on previous results, imidazolium, ammonium, bifunctional, and other ionic liquids have high potential as extraction agents for REEs with high extraction efficiencies, replacing conventional extractants due to sustainability requirements [47,48]. Surprisingly, Aliquat 336(NO_3) did not give the expected REE extraction results, with the extraction efficiency values being extremely low in GA solutions, except for La (~80%) and Ce (~24%).

Table 1. Extraction efficiencies (*E* %) of REEs from graptolite-argillite using different extraction agents (marked in Table and text) from 3 M HNO_3 aqueous solutions.

	E %															
	Sc	±	**Y**	±	**La**	±	**Ce**	±	**Pr**	±	**Nd**	±	**Sm**	±	**Eu**	±
EMIMDEPO4	99.5	1.0	99.5	0.9	99.8	1.1	99.6	0.5	99.5	0.3	99.6	0.7	99.6	0.8	99.6	1.5
Aliquat336	0	0	0	0	79.6	3.6	24.2	2.1	0	0	0	0	1.4	0.2	0	0
TBP	0	0	0	0	0	0	0	0	0	0	0	0	4.1	0.5	0	0
	Gd	±	**Tb**	±	**Dy**	±	**Ho**	±	**Er**	±	**Tm**	±	**Yb**	±	**Lu**	±
EMIMDEPO4	99.6	1.5	99.6	2	99.6	0.8	99.5	0.9	99.5	0.7	99.5	1.2	99.4	2	99.3	1.2
Aliquat336	2.3	0.1	0	0	0	0	0.7	0.1	1.7	0.1	0	0	1.6	0.1	0	0
TBP	0	0	0	0	0	0	0	0	0	0	0	0	0	0	0	0

The extraction efficiencies calculated for the PH from the ICP-MS/MS results indicate that the concentration of HCl significantly influences the extraction profile using D2EHPA as an extraction agent (Table 2). The extraction efficiencies for Aliquat 336(NO_3) were low, and the results are not included in the analysis of this paper. Experimental data for D2EHPA show that in more concentrated HCl solutions (5 M and 3 M), HREE and Sc can be extracted selectively. In contrast, LREEs (La-Sm), Y, Eu, Gd, and Tb are not extracted under these conditions. This is in good agreement with previous observations that D2EHPA has better selectivity of HREE over LREE, which can be attributed to the higher charge density of HREE [49]. The LREEs, Y, Eu, Gd, Tb, and Dy can be extracted more effectively using lower HCl (2 M and 1 M) concentrations (Table 2). It is important to note that the extraction efficiency of Y increases significantly at lower concentrations, but the extraction efficiency of Sc is higher in a more acidic environment. We propose that for the extraction of LREE (La-Sm), the concentration of HCl needs to be lower than 1 M. Still, so far, our results indicate that leaching at less acidic concentrations becomes problematic. Therefore, leaching in a more acidic environment, followed by partial neutralization of leachate, might be an option. It can be concluded that extraction in HCl media using different acid concentrations shows promising results for separating HREE and Sc from LREE.

Table 2. Extraction efficiencies of REEs from phosphorite ore (Ülgase) at different hydrochloric acid molar concentrations using D2EHPA as an extractant.

	E %															
	Sc	±	**Y**	±	**La**	±	**Ce**	±	**Pr**	±	**Nd**	±	**Sm**	±	**Eu**	±
5M HCl	98.08	1.0	16.55	2.9	−4.89	1.1	−4.62	0.5	−2.96	0.3	−3.01	0.7	−4.06	0.8	−4.21	1.5
3M HCl	75.57	2.3	13.39	2.3	−6.92	3.6	−6.98	2.1	−4.72	0	−6.59	0	−5.48	0.2	5.4	0
2M HCl	67.02	3.3	39.04	1.9	4.18	0.2	1.97	0.1	1.27	0.1	3.22	0.2	4.29	0.2	4.08	0.2
1M HCl	17.57	1.7	92.07	3.5	0.38	0	2.64	0	3.43	0	3.78	0	18.53	0.5	29.11	0
	Gd	±	**Tb**	±	**Dy**	±	**Ho**	±	**Er**	±	**Tm**	±	**Yb**	±	**Lu**	±
5M HCl	−2.33	1.5	−0.42	2	3.35	0.8	9.63	0.9	20.79	0.7	44.16	1.2	65.61	2	72.33	1.2
3M HCl	−3.88	0.1	−1.05	0	4.93	0	7.6	0.1	14.28	0.1	32.22	0	44.34	0.1	43.27	0
2M HCl	5.27	0	12.64	0	21.63	0	26.8	0	36.39	0	59.44	0	50.08	0	36.72	0
1M HCl	30.59	1.5	59.99	2.9	79.35	3.9	84.59	4.2	82.39	4.1	92.71	4.6	47.4	2.3	67.6	3.4

The extraction results of the PH (Iru) acidic solutions with D2EHPA are highly dependent on the concentration of HNO_3 (Table 3), similar to the HCl results. High extraction efficiency values were calculated for all REEs from dilute nitric acid solution with D2EHPA. However, the extraction efficiency values from the concentrated nitric acid solution are better for HREE and negligible for LREE.

Table 3. Extraction efficiencies of phosphorite ore (Iru) at different nitric acid molar concentrations using D2EHPA as an extractant.

	E %															
	Sc	±	**Y**	±	**La**	±	**Ce**	±	**Pr**	±	**Nd**	±	**Sm**	±	**Eu**	±
0.03 M HNO_3	99.30	1.00	99.57	1.00	92.98	1.00	94.51	1.00	94.91	1.00	95.27	1.00	97.39	1.00	98.01	1.00
7.5 M HNO_3	92.86	1.00	27.06	2.7	2.60	0.26	3.27	0.32	6.61	0.66	3.52	0.35	4.80	0.48	5.27	0.52
	Gd	±	**Tb**	±	**Dy**	±	**Ho**	±	**Er**	±	**Tm**	±	**Yb**	±	**Lu**	±
0.03 M HNO_3	97.77	1.00	98.81	1.00	99.24	1.00	99.53	1.00	99.50	1.00	99.55	1.00	99.44	1.00	99.62	1.00
7.5 M HNO_3	4.60	0.46	8.02	0.80	14.12	1.41	19.10	1.91	27.70	2.77	43.17	4.31	58.38	5.38	66.38	6.63

It is essential to point out that the extraction efficiency of Sc is not influenced by the acidic concentration in nitric acid media and remains higher than 90%, which is different from hydrochloric acid media. Thus, for the selective extraction of HREE with D2EHPA, more concentrated HCl and HNO_3 should be used. All the results presented in this paper are promising, mainly if ionic liquid extraction has been conducted. However, additional systematic extraction and separation should be conducted to selectively separate REE and DRM from Estonian GA and PH ore samples.

4. Conclusions

The leaching process of Estonian GA and PH was thoroughly studied using HCl and HNO_3 with different concentrations, including *Aqua Regia*. A higher concentration of REEs in GA and PH was achieved in more concentrated acid solutions, especially in HNO_3. A variation in phosphate-based extractants and ILs was explored for the extraction process of REEs. The extraction efficiencies in GA complex systems are highest when EMImDEPO4 is used as an extraction agent. In PH systems, more concentrated HCl and HNO_3 solutions gave better results for the selective extraction of HREEs from LREEs.

Continuous and systematic analysis of different leaching and extraction agents should be conducted to extract and separate REEs from Estonian GA and PH successfully.

Supplementary Materials: The following supporting information can be downloaded at: https://www.mdpi.com/article/10.3390/met15060608/s1.

Author Contributions: Conceptualization: S.J., L.S. and O.O.; methodology: S.J. and L.S.; software: C.S. and O.O.; validation: S.J. and C.S.; formal analysis: L.S.; investigation; S.J. and C.S.; resources: E.L.; data curation: L.S. and O.O.; writing—original draft: C.S. and S.J.; preparation: L.S. and O.O.; writing—review and editing: E.L., L.S., O.O. and S.J.; visualization: O.O.; supervision: E.L.; project administration, E.L.; funding acquisition, E.L. All authors have read and agreed to the published version of the manuscript.

Funding: This work was supported by the Estonian Research Council Grant PRG676 and PRG2677; the Centre of Nanomaterials Technologies and Research project TT13 (NAMUR+); Estonian Ministry of Education and Research project TK210 "Centre of Excellence in Sustainable Green Hydrogen and Energy Technologies" (01.01.2024–31.12.2030); and by the projects "Increasing the knowledge intensity of Ida-Viru entrepreneurship" (ÕÜF12, ÕÜF13) co-funded by the European Union.

Data Availability Statement: The original contributions presented in this study are included in the article/Supplementary Material. Further inquiries can be directed to the corresponding author.

Acknowledgments: Päärn Paiste and Liis Vitsut for contribution to this work.

Conflicts of Interest: The authors declare no conflict of interest.

References

1. Dushyantha, N.; Batapola, N.; Ilankoon, I.M.S.K.; Rohitha, S.; Premasiri, R.; Abeysinghe, B.; Ratnayake, N.; Dissanayake, K. The story of rare earth elements (REEs): Occurrences, global distribution, genesis, geology, mineralogy and global production. *Ore Geol. Rev.* **2020**, *122*, 103521. [CrossRef]
2. Kanazawa, Y.; Kamitani, M. Rare earth minerals and resources in the world. *J. Alloys Compd.* **2006**, *408–412*, 1339–1343. [CrossRef]
3. McLennan, S.M.; Ross Taylor, S. Geology, Geochemistry and Natural Abundances. In *Encyclopedia of Inorganic and Bioinorganic Chemistry*; John Wiley & Sons, Ltd.: Hoboken, NJ, USA, 2012; ISBN 978-1-119-95143-8.
4. Jordens, A.; Cheng, Y.P.; Waters, K.E. A review of the beneficiation of rare earth element bearing minerals. *Miner. Eng.* **2013**, *41*, 97–114. [CrossRef]
5. Chen, Z.; Li, Z.; Chen, J.; Kallem, P.; Banat, F.; Qiu, H. Recent advances in selective separation technologies of rare earth elements: A review. *J. Environ. Chem. Eng.* **2022**, *10*, 107104. [CrossRef]
6. Ilankoon, I.M.S.K.; Dushyantha, N.P.; Mancheri, N.; Edirisinghe, P.M.; Neethling, S.J.; Ratnayake, N.P.; Rohitha, L.P.S.; Dissanayake, D.M.D.O.K.; Premasiri, H.M.R.; Abeysinghe, A.M.K.B.; et al. Constraints to rare earth elements supply diversification: Evidence from an industry survey. *J. Clean. Prod.* **2022**, *331*, 129932. [CrossRef]
7. Zhou, B.; Li, Z.; Chen, C. Global Potential of Rare Earth Resources and Rare Earth Demand from Clean Technologies. *Minerals* **2017**, *7*, 203. [CrossRef]
8. Tukker, A. Rare Earth Elements Supply Restrictions: Market Failures, Not Scarcity, Hamper Their Current Use in High-Tech Applications. *Environ. Sci. Technol.* **2014**, *48*, 9973–9974. [CrossRef]
9. Coey, J.M.D. Perspective and Prospects for Rare Earth Permanent Magnets. *Engineering* **2020**, *6*, 119–131. [CrossRef]

10. Podmiljšak, B.; Saje, B.; Jenuš, P.; Tomše, T.; Kobe, S.; Žužek, K.; Šturm, S. The Future of Permanent-Magnet-Based Electric Motors: How Will Rare Earths Affect Electrification? *Materials* **2024**, *17*, 848. [CrossRef]
11. Bailey, G.; Mancheri, N.; Van Acker, K. Sustainability of Permanent Rare Earth Magnet Motors in (H)EV Industry. *J. Sustain. Metall.* **2017**, *3*, 611–626. [CrossRef]
12. Fernandez, V. Rare-earth elements market: A historical and financial perspective. *Resour. Policy* **2017**, *53*, 26–45. [CrossRef]
13. Zhao, S.; Wang, P.; Chen, W.; Wang, L.; Wang, Q.-C.; Chen, W.-Q. Supply and demand conflicts of critical heavy rare earth element: Lessons from gadolinium. *Resour. Conserv. Recycl.* **2023**, *199*, 107254. [CrossRef]
14. Tokimatsu, K.; Murakami, S.; Adachi, T.; Ii, R.; Yasuoka, R.; Nishio, M. Long-term demand and supply of non-ferrous mineral resources by a mineral balance model. *Miner. Econ.* **2017**, *30*, 193–206. [CrossRef]
15. Voolma, M.; Soesoo, A.; Hade, S.; Hints, R.; Kallaste, T. Geochemical heterogeneity of estonian graptolite argillite. *Oil Shale* **2013**, *30*, 377. [CrossRef]
16. Kaljuvee, T.; Tõnsuaadu, K.; Einard, M.; Mikli, V.; Kivimäe, E.-K.; Kallaste, T.; Trikkel, A. Thermal Behavior of Estonian Graptolite–Argillite from Different Deposits. *Processes* **2022**, *10*, 1986. [CrossRef]
17. Vind, J.; Tamm, K. Review of the extraction of key metallic values from black shales in relation to their geological and mineralogical properties. *Miner. Eng.* **2021**, *174*, 107271. [CrossRef]
18. Vind, J.; Ofili, S.; Mänd, K.; Soesoo, A.; Kirsimäe, K. Redox-sensitive trace metal hyper-enrichment in Tremadocian Alum Shale (graptolite argillite) in northwestern Estonia, Baltic Palaeobasin. *Chem. Geol.* **2023**, *640*, 121746. [CrossRef]
19. Menert, A.; Korb, T.; Orupõld, K.; Teemusk, A.; Sepp, H.; Mander, Ü.; Ilmjärv, T.; Truu, J.; Paiste, P.; Kirsimäe, K.; et al. Methanogenesis and metal leaching on anaerobic decomposition of graptolite argillite. *Environ. Technol. Innov.* **2023**, *31*, 103139. [CrossRef]
20. Yang, X.; Tamm, K.; Piir, I.; Kuusik, R.; Trikkel, A.; Tõnsuaadu, K. Evaluation of Estonian phosphate rock by flotation. *Miner. Eng.* **2021**, *171*, 107127. [CrossRef]
21. Brahim, J.A.; Hak, S.A.; Achiou, B.; Boulif, R.; Beniazza, R.; Benhida, R. Kinetics and mechanisms of leaching of rare earth elements from secondary resources. *Miner. Eng.* **2022**, *177*, 107351. [CrossRef]
22. Jha, M.K.; Kumari, A.; Panda, R.; Kumar, J.R.; Yoo, K.; Lee, J.Y. Review on hydrometallurgical recovery of rare earth metals. *Hydrometallurgy* **2016**, *165*, 2–26. [CrossRef]
23. Peelman, S.; Sun, Z.H.I.; Sietsma, J.; Yang, Y. Hydrometallurgical Extraction of Rare Earth Elements from Low Grade Mine Tailings. In *Rare Metal Technology 2016*; John Wiley & Sons, Ltd.: Hoboken, NJ, USA, 2016; pp. 17–29, ISBN 978-1-119-27483-4.
24. Costis, S.; Mueller, K.K.; Coudert, L.; Neculita, C.M.; Reynier, N.; Blais, J.-F. Recovery potential of rare earth elements from mining and industrial residues: A review and cases studies. *J. Geochem. Explor.* **2021**, *221*, 106699. [CrossRef]
25. El-Nadi, Y.A.; El-Hefny, N.E.; Aly, H.F. Solvent extraction and recovery of Y(III) and Yb(III) from fluorspar mineral. *Int. J. Miner. Metall. Mater.* **2013**, *20*, 713–719. [CrossRef]
26. Liu, T.; Chen, J. Extraction and separation of heavy rare earth elements: A review. *Sep. Purif. Technol.* **2021**, *276*, 119263. [CrossRef]
27. Bandara, H.M.D.; Field, K.D.; Emmert, M.H. Rare earth recovery from end-of-life motors employing green chemistry design principles. *Green Chem.* **2016**, *18*, 753–759. [CrossRef]
28. Arrachart, G.; Couturier, J.; Dourdain, S.; Levard, C.; Pellet-Rostaing, S. Recovery of Rare Earth Elements (REEs) Using Ionic Solvents. *Processes* **2021**, *9*, 1202. [CrossRef]
29. Wang, K.; Adidharma, H.; Radosz, M.; Wan, P.; Xu, X.; Russell, C.K.; Tian, H.; Fan, M.; Yu, J. Recovery of rare earth elements with ionic liquids. *Green Chem.* **2017**, *19*, 4469–4493. [CrossRef]
30. Baba, Y.; Kubota, F.; Kamiya, N.; Goto, M. Kinetics and mechanisms of leaching of rare earth elements from secondary resources. *J. Chem. Eng. Jpn.* **2011**, *44*, 679–685. [CrossRef]
31. Soesoo, A. Main Precambrian and Paleozoic Mineral Resources of Estonia. *Asp. Min. Miner. Sci.* **2021**, *6*, 729–732. [CrossRef]
32. Soesoo, A.; Vind, J.; Hade, S. Uranium and Thorium Resources of Estonia. *Minerals* **2020**, *10*, 798. [CrossRef]
33. Hade, S.; Soesoo, A. Estonian graptolite argillites revisited: A future resource? *Oil Shale* **2014**, *31*, 4. [CrossRef]
34. Zheng, Q.; Zhang, Y.; Liu, T.; Huang, J.; Xue, N. Vanadium extraction from black shale: Enhanced leaching due to fluoride addition. *Hydrometallurgy* **2019**, *187*, 141–148. [CrossRef]
35. Sevim, F.; Saraç, H.; Kocakerim, M.M.; Yartaşı, A. Dissolution Kinetics of Phosphate Ore in H_2SO_4 Solutions. *Ind. Eng. Chem. Res.* **2003**, *42*, 2052–2057. [CrossRef]
36. Clavier, N.; Podor, R.; Dacheux, N. Crystal chemistry of the monazite structure. *J. Eur. Ceram. Soc.* **2011**, *31*, 941–976. [CrossRef]
37. Senanayake, G. Gold leaching by copper(II) in ammoniacal thiosulphate solutions in the presence of additives. Part I: A review of the effect of hard–soft and Lewis acid-base properties and interactions of ions. *Hydrometallurgy* **2012**, *115–116*, 21–29. [CrossRef]
38. Watling, H.R. Review of Biohydrometallurgical Metals Extraction from Polymetallic Mineral Resources. *Minerals* **2014**, *5*, 1–60. [CrossRef]
39. Zhu, X.; Wang, Z.; Chen, H. Advances in Isotope Geochronology and Isotope Geochemistry: A Preface. *J. Earth Sci.* **2022**, *33*, 1–4. [CrossRef]

40. Waste and Recycling. Available online: https://environment.ec.europa.eu/topics/waste-and-recycling_en (accessed on 18 November 2022).
41. Jürjo, S.; Oll, O.; Paiste, P.; Külaviir, M.; Zhao, J.; Lust, E. Electrochemical co-reduction of praseodymium and bismuth from 1-butyl-1-methylpyrrolidinium bis (fluorosulfonyl) imide ionic liquid. *Electrochem. Commun.* **2022**, *138*, 107285. [CrossRef]
42. Jürjo, S.; Oll, O.; Lust, E. Yttrium Separation from Phosphorite Extract Using Liquid Extraction with Room Temperature Ionic Liquids Followed by Electrochemical Reduction. *Metals* **2024**, *14*, 927. [CrossRef]
43. Jürjo, S.; Siinor, L.; Siimenson, C.; Paiste, P.; Lust, E. Two-Step Solvent Extraction of Radioactive Elements and Rare Earths from Estonian Phosphorite Ore Using Nitrated Aliquat 336 and Bis(2-ethylhexyl) Phosphate. *Minerals* **2021**, *11*, 388. [CrossRef]
44. Sepúlveda, R.; Toro, N.; Hernández, P.; Navarro, P.; Vargas, C.; Gálvez, E.; Castillo, J. Solvent Extraction of Metal Ions from Synthetic Copper Leaching Solution Using R4NCy. *Metals* **2022**, *12*, 1053. [CrossRef]
45. Gorzin, H.; Ghaemi, A.; Hemmati, A.; Maleki, A. Studies on effective interaction parameters in extraction of Pr and Nd using Aliquat 336 from NdFeB magnet-leaching solution: Multiple response optimizations by desirability function. *J. Mol. Liq.* **2021**, *324*, 115123. [CrossRef]
46. Cheremisina, O.; Ponomareva, M.; Sergeev, V.; Mashukova, Y.; Balandinsky, D. Extraction of Rare Earth Metals by Solid-Phase Extractants from Phosphoric Acid Solution. *Metals* **2021**, *11*, 991. [CrossRef]
47. Quijada-Maldonado, E.; Romero, J. Solvent extraction of rare-earth elements with ionic liquids: Toward a selective and sustainable extraction of these valuable elements. *Curr. Opin. Green Sustain. Chem.* **2021**, *27*, 100428. [CrossRef]
48. Kaim, V.; Rintala, J.; He, C. Selective recovery of rare earth elements from e-waste via ionic liquid extraction: A review. *Sep. Purif. Technol.* **2023**, *306*, 122699. [CrossRef]
49. Gergoric, M.; Ekberg, C.; Steenari, B.-M.; Retegan, T. Separation of Heavy Rare-Earth Elements from Light Rare-Earth Elements via Solvent Extraction from a Neodymium Magnet Leachate and the Effects of Diluents. *J. Sustain. Metall.* **2017**, *3*, 601–610. [CrossRef]

metals

MDPI

Article

Cu(II) and Ni(II) Adsorption on Torrefied Wood Waste Biomass

Marjana Simonič *, Darko Goričanec, Aleksandra Petrovič, Ilda Silić and Danijela Urbancl

Faculty of Chemistry and Chemical Engineering, University of Maribor, Smetanova 17, 2000 Maribor, Slovenia; darko.goricanec@um.si (D.G.); aleksandra.petrovic@um.si (A.P.); ilda.silic@student.um.si (I.S.); danijela.urbancl@um.si (D.U.)

* Correspondence: marjana.simonic@um.si; Tel.: +386-2-2294-472

Abstract: The aim of the research was to study the torrefaction processes of wood biomass, compare the product characteristics at different torrefaction temperatures, and assess both moisture adsorption on raw and torrefied samples, as well as metal (Cu(II) and Ni(II)) adsorption on torrefied biomass. The novelty of the research was to investigate whether the presence of adsorbed metals in torrefied biomass significantly affects the energetic properties of the torrefied biomass, compared to torrefied biomass without metals. First, wood samples were torrefied at temperatures of 250 °C, 350 °C, and 400 °C. Following torrefaction, thermogravimetric analysis (TGA) was performed to evaluate mass loss and thermal stability. Next, changes in surface functional groups were examined, and higher heating values (*HHV*) were measured to assess the energy content. The results showed that torrefaction significantly increased the hydrophobicity of the biomass, leading to reduced moisture adsorption and enhanced material properties. Additionally, the adsorption of Cu(II) and Ni(II) ions on torrefied biomass was investigated. The results showed that the adsorption efficiency for Cu(II) was higher, reaching 62.4%, compared to Ni(II) at 21.2%. The adsorption process followed a pseudo-second-order kinetic model, which indicated that chemisorption was the dominant mechanism.

Keywords: adsorption; torrefaction; nitrogen atmosphere; metals

1. Introduction

Adsorption is the process in which a substance (gas, liquid, or solid) binds to the surface of another substance. It involves two components: the adsorbate and the adsorbent. The adsorbate is the substance that binds to the surface of the adsorbent, such as gases like H_2, N_2, and O_2, or materials like charcoal, silicone gel, and aluminum oxide [1]. By altering the properties of the liquid phase, such as concentration, temperature, or pH, adsorbed substances can be released back from the surface into the liquid phase in a process known as desorption [2]. Adsorption can be classified into two types: physical and chemical. If the attractive forces between the adsorbate and the adsorbent result from chemical bonding, the process is called chemical adsorption or chemisorption. If the interaction is physical, it is referred to as physical adsorption or physisorption. Adsorption has a wide range of applications, including in metallurgy, ion exchange, water treatment, and moisture removal. The adsorption process can be modeled using various kinetic models, including the pseudo-first-order reaction, pseudo-second-order reaction, Weber–Morris model, and the Elovich model [3].

The presence of Cu(II) in water is a concern due to its common occurrence in various industrial processes, while the electroplating industry is one of the major sources of Ni(II) pollution [4]. Both Cu(II) and Ni(II) can pose significant health risks at elevated

https://doi.org/10.3390/met15030304

concentrations, with the WHO setting tolerable levels at 2 mg/L and 20 μg/L, respectively, for drinking water [5]. Interestingly, leachate residues from processing deep-sea nodules have been found to be effective adsorbents for heavy metals like Cu(II) and Ni(II). However, adsorption is most effective when carried out at normal temperatures. In mixed systems containing multiple metals, adsorption capacity tends to be lower compared to systems with individual metals. Additionally, the high concentration of ions like K^+, Na^+, Ca^{2+}, and Mg^{2+} in water can compete with metal cations, reducing the efficiency of metal ion adsorption [3].

Biomass, while a renewable energy source, presents several challenges due to its high moisture and oxygen content, low energy density, hydrophilic nature, and heterogeneity. These characteristics lead to high energy consumption during grinding and pelletizing, low combustion temperatures, inefficient combustion, and the production of smoke and flue gases. The hydrophilic nature of biomass also creates storage issues as it is prone to biodegradation. These disadvantages impact its energy density and stability. Therefore, pretreatment methods are required to enhance its properties and make biomass more efficient for use. The conversion of biomass has garnered increasing attention from researchers due to the significant improvements in its properties when compared to raw biomass [6]. Among the various methods, torrefaction has emerged as an effective process.

Torrefaction is a controlled carbonization process that transforms biomass into a product with properties similar to coal. This process involves heating biomass to temperatures ranging from 200 °C to 300 °C in an oxygen-free environment or in the presence of minimal oxygen. Nitrogen is typically used as a carrier gas to maintain a non-oxidizing atmosphere [7]. During torrefaction, hemicellulose decomposes, releasing water, CO_2, CO, and various organic acids. As a result, the mass of the biomass is reduced by approximately 30%, and its energy density or heating value is decreased by around 20% [8]. Thermal methods at temperatures below 200 °C are generally used for wood preservation, while higher temperatures, as in torrefaction, are used for energy purposes.

In addition to temperature, the duration of the torrefaction process plays an important role as it can last anywhere from a few minutes to several hours. Torrefaction increases the hydrophobicity of biomass, making it more resistant to moisture adsorption and improving storage control. The reduction in water content also makes biomass lighter and less prone to decomposition [9]. With these enhanced properties, torrefied biomass requires less storage capacity and has the potential to reduce transport costs.

The research aimed to highlight the improved properties of torrefied biomass compared to raw wood, particularly in terms of hydrophobicity, energy content, and adsorption capacity. The study was divided into three main objectives:

1. To examine the effect of temperature on the properties of torrefied biomass, with a focus on moisture adsorption differences between raw biomass and torrefied biomass at temperatures of 250, 350, and 400 °C. The study investigated how torrefaction influences mass loss, thermal stability, functional group composition, and higher heating values (*HHV*).
2. To evaluate the hydrophobicity of torrefied biomass and identify the materials with the best adsorption properties. These materials were then tested for their ability to adsorb metals such as copper (Cu(II)) and nickel (Ni(II)) in model solutions.
3. To assess the potential of torrefied biomass with adsorbed metals as an energy source. Specifically, the study investigated whether the presence of bound metals significantly affects the energy properties of the material, especially its higher calorific value.

2. Materials and Methods

2.1. The Torreefaction Process and Analyses

The whole torrefaction process involves five steps:

1. Initial heating: Biomass is heated up to 100 °C.
2. Pre-drying: The biomass is heated at a constant temperature of 100 °C to evaporate free water, continuing until a stable weight is achieved.
3. Post-drying and intermediate heating: The biomass temperature rises to 200 °C. Weight loss can occur during this phase.
4. Torrefaction: This phase begins when the temperature reaches 200 °C and ends when the biomass cools down again from a certain temperature to 200 °C. The torrefaction temperature is defined as the highest constant temperature. During this period, most of the biomass mass is lost.
5. Cooling: The torrefied product is additionally cooled to the desired final temperature [10].

The torrefaction process was conducted using a Gradient Tube Furnace (Carbolite Gero, Bubsheim„Germany) tube furnace equipped with a removable trough in the working tube. Sensors were placed on the working tube to monitor the temperature in each chamber (Figure 1). The samples underwent torrefaction at three different temperatures: 250 °C, 350 °C and 400 °C. The torrefaction was performed in a nitrogen atmosphere (N_2, 3 L/min). The inlet gas was preheated to 50 °C. The duration of torrefaction under N_2 atmosphere was determined based on the initial gas composition, whereas during the torrefaction process, gases are produced, and their concentration was analyzed with an ABB AO2020 gas analyzer (ABB, Zurich, Switzerland). The volume fractions of CO_2, CO, CH_4, and oxygen were measured. A cooled impinger train was installed upstream of the gas analyzer to ensure that the gas remains tar-free and dry.

Figure 1. Schematic presentation of the process.

The mass loss W_L was calculated following Equation (1):

$$W_L = \frac{m_0 - m_k}{m_0} \cdot 100 \quad (1)$$

where

W_L—mass loss (%);

m_0—initial sample mass (g);

m_k—final sample mass (g).

The ratio of energy density (*EDR*) and efficiency of energy *EY* was calculated according to Equations (2) and (3) [11]:

$$\text{EDR} = \frac{HHV\ tor}{HHVs} \quad (2)$$

where

EDR—the ratio of energy density (%);

HHVtor—*HHV* of torrefied sample (MJ/kg);
HHVs—*HHV* raw sample (MJ/kg).

$$EY = EDR \cdot MY \tag{3}$$

where

EY—efficiency of energy (%).

Thermogravimetric (TGA) and derivative thermogravimetric (DTG) analyses were performed. The TGA/DTG measurements were performed using Mettler Toledo TGA/DSC^{3+} (Mettler Toledo, Ljubljana, Slovenia) in oxygen atmosphere. The measurement was set as follows: the sample was heated from 30 °C to 105 °C with 20 K/min rate, then the heating was at a constant temperature of 105 °C for 5 min. Then, the temperature increased from 105 °C to 550 °C at 25 K/min rate. At 550 °C, the sample was heated for 20 min. Finally, the temperature was increased to 815 °C, and the sample was heated for 20 min at 25 K/min rate.

Infrared spectroscopy with Fourier transform (FTIR) was used to record the infrared absorption spectrum of a substance, and Fourier transform was used for data processing. The FTIR was determined using the NICOLET iS50 FT-IR apparatus (Omega, Ljubljana, Slovenia). The spectra of the samples were recorded using the classic ATR method, namely in the range from 400 cm^{-1} to 4000 cm^{-1}.

The higher heating value (*HHV*) was determined using the IKA Isoperibol 6000 bomb calorimeter (Kefo, Ljubljana, Slovenia) according to the standardized method.

The Modular Humidity Generator was attached to Mettler Toledo TGA/DSC^{3+} for humidity control. Moisture adsorption was performed with four different samples, which included raw wood, as well as torrefied wood, at 250 °C, 350 °C, and 400 °C. Four measurement segments were set. The relative humidity was measured up to 30% for 90 min in the first segment, from 30% to 45% for 90 min in the second segment, from 60% for 90 min in the third segment, and up to 75% for 90 min in the last one. Moisture adsorption was performed at 30 °C 100 mL/min flow in nitrogen atmosphere.

2.2. The Adsorption of Cu(II) and Ni(II)

The sample used was wood torrefied at 400 °C, and adsorption of two metals, Cu(II) and Ni(II), was performed.

Two stock solutions of Cu(II) and Ni(II) ions (with concentration of 1 g/L) were obtained by dissolving exact amounts of high-purity $CuSO_4$ (Merck, Darmstadt, Germany) and high-purity $NiSO_4$ (Merck, Darmstadt, Germany) in 1 L of distilled water, respectively. The working solutions of 100 mL were obtained by diluting the appropriate amount of each stock solution with the Millipore water sample, in which the initial mass of adsorbent varied between 1 and 0.1 g (1 g, 0.5 g, 0.2 g in 0.1 g). The adsorption experiments were performed in 300-mL Erlenmeyer flasks with 100 mL of working solutions stirred by a magnetic stirrer at 250 rpm. All tests were conducted at room temperature (22.0 ± 1.0).

The samples for the analysis of metal content were collected at the beginning of the adsorption and then at different time intervals, of 30, 60, 120, and 1440 min, when equilibrium was reached. pH value change was followed. Prior to the analysis, the samples were filtered through glass fiber filters (0.45 μm). The metal content in the solutions was subsequently analyzed using a photometer PF-12 PLUS (Micro+Polo, Maribor, Slovenia).

The effect of contact time on the biosorption of Cu(II) ions caused by torrefied material was studied at an initial metal concentration of 50 mg/L (pH = 5), and the contact time increased from 0 min up to 180 min, based on the literature [12].

In our study, pseudo-first-order reaction and pseudo-second-order reaction were studied. Equation (4) shows pseudo-first-order reaction, while Equation (5) shows model's pseudo-second-order reaction.

$$\ln\left(q_e - q_t\right) = \ln q_e - k_1 \cdot t \quad (4)$$

where

q_e—amount of adsorbed matter in equilibrium (mg/g);
q_t—amount of adsorbed matter at time t (mg/g);
k_1—rate constant of pseudo-first-order reaction (1/min);
t—time (min).

$$\frac{t}{q_t} = \frac{1}{k_2 \cdot q_e^2} + \frac{t}{q_e} \quad (5)$$

where

k_2—rate constant at pseudo-second-order reaction (g/(mg·min));

Equations (6) and (7) show the calculation of adsorbed amount q_t (mg/g) and adsorption efficiency E (%).

$$q_t = \frac{c_0 - c_t}{m_{tore}} \cdot V \quad (6)$$

where

c_0—initial concentration (mg/L);
c_t—concentration at time t (mg/L);
m_{tore}—mass of torrefied material (g);
V—volume of solution (mL).

$$E = \frac{c_0 - c_t}{c_0} \cdot 100 \quad (7)$$

Factors affecting metal adsorption are initial concentration, pH, temperature, and coexisting ions. The influence of the initial concentration is shown in a certain concentration range, where the adsorption capacity of heavy metals increases with the increase in the initial concentration until the carbon adsorbent reaches saturation. At lower pH values below 6, metal ions are usually in the cationic form, while at higher pH values above 9 they may appear in the anionic form. When the pH value is lower than the point of zero charge value where the zeta potential of adsorbents is equal to zero, there is a greater exchange of H^+ ions with metal cations and an increase in the number of positive charges on the surface of carbon materials. In our previous work, the optimum pH value for Cu(II) and Ni(II) adsorption was determined to be pH = 5 [12].

After the adsorption, the samples of torrefied biomass together with the adsorbed Cu(II) and Ni(II) ions were torrefied once again in order to assess the energy properties of torrefied biomass with adsorbed metals compared with those of torrefied material without adsorbed metals.

3. Results

3.1. Torrefaction Process Results

Torrefaction of wood material was performed in the N_2 atmosphere at temperatures of 250 °C, 350 °C, and 400 °C. The samples are seen in Figure 2. As the temperature increased, the color of torrefied material changed from light to dark, reflecting alternations in its chemical composition. Several chemical reactions occurred during torrefaction, including the degradation of the cellulose, hemicellulose, oxidation, and reduction processes of lignin, and the enzymatically controlled Maillard reaction [13]. The interaction between proteins and polysaccharides through the Maillard reaction can alter the structure of proteins and enhance the functional properties of the material. The oxidative reaction of extracts in wood contributed to the darkening of the material, indicating the progression of carbonization on the wood's surface.

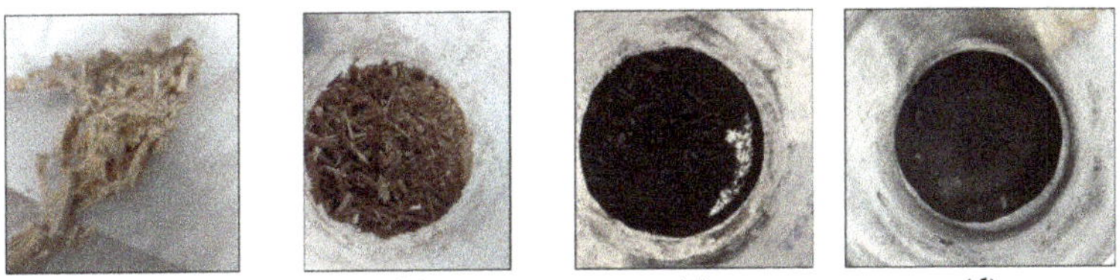

(**a**) (**b**) (**c**) (**d**)

Figure 2. Samples of wood: (**a**) raw; (**b**) torrefied at 250 °C; (**c**) torrefied at 300 °C; (**d**) torrefied at 400 °C.

Figure 3 shows the gas analyses of torrefied material at different temperatures in N_2 atmosphere with a flow rate of 3 L/min. The analyses of material torrefied up to 250 °C showed the highest production of CO_2 (2.2%), while the oxygen release peaked at 1.5% after 53.8 min. Methane levels remained stable at 0.4%, and CO reached the maximum of 0.2%. In the material torrefied up to 350 °C, no oxygen was detected, while CO_2 reached a maximum at 20.8%, and CO reached a peak of 7.5%. The analyses of material torrefied up to 400 °C showed that the CO_2 remained the dominant gas at 14.8%, oxygen was not detected, methane reached 1.2%, and CO peaked at 6.0%.

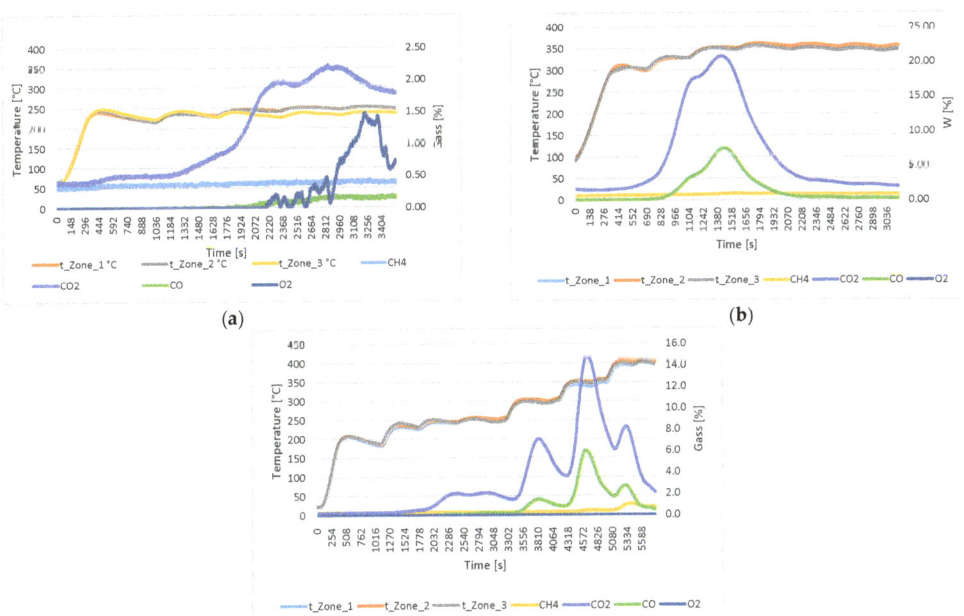

(**c**)

Figure 3. Gas analyses of wood material in N_2 atmosphere with 3 L/min flow at the following temperatures: (**a**) 250 °C; (**b**) 350 °C; (**c**) 400 °C.

The results of mass loss (*WL*) at 250 °C, 350 °C, and 400 °C are shown in Figure 4.

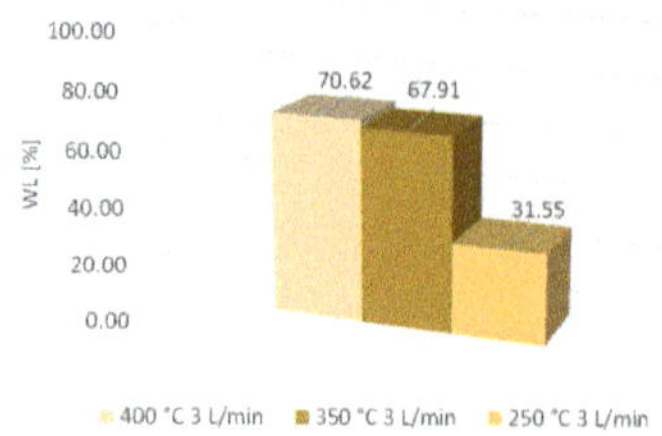

Figure 4. Mass losses at 250 °C, 350 °C, and 400 °C.

As the sample mass increased, the mass loss also increased. The difference between mass losses at temperatures of 350 °C and 400 °C was very low; it was determined to be 2.7%. The difference between mass losses at 250 °C and 350 °C was much higher; it was determined to be 36.4%. Therefore, the temperature increased up to 350 °C and contributed to higher activity, while the decarbonization progress was higher compared to the lower temperature of 250 °C. The cellulose, as well as the hemicellulose, started to degrade at a temperature above 270 °C, and consequently, mass loss was higher at 350 °C [11].

The TGA analysis results of six samples are shown in Figure 5. TG and DTG curves are seen in Figure 5. *Ti* represents the temperature of initial sample degradation, and *Tb* the temperature at the highest degradation rate.

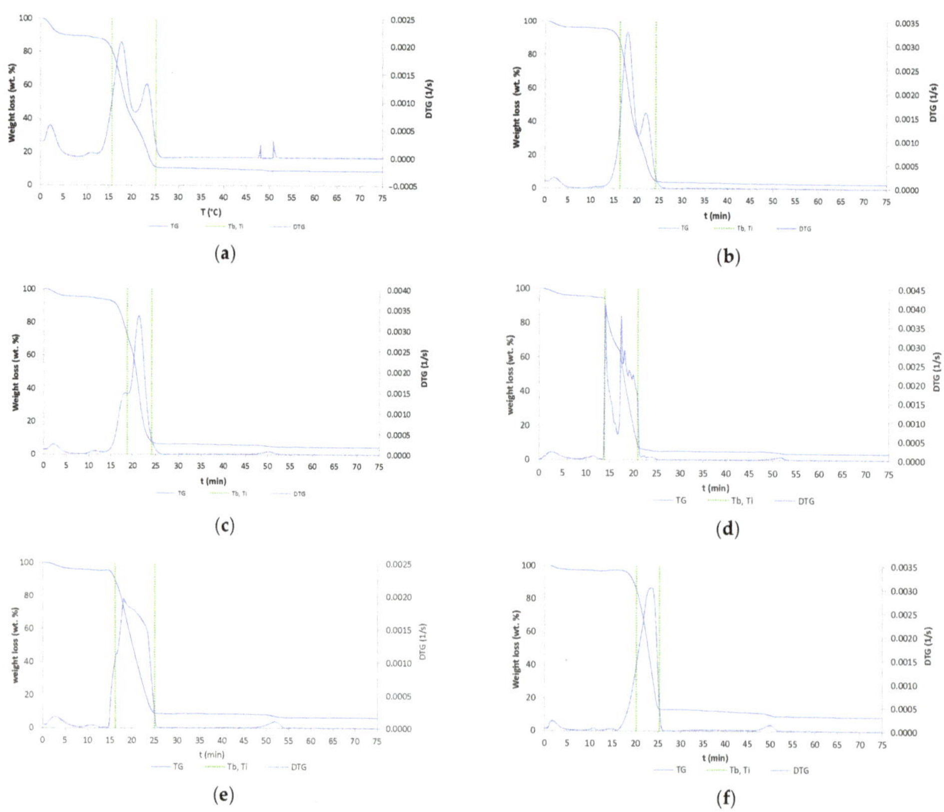

Figure 5. TG and DTG curves: (**a**) raw wood; (**b**) wood torrefied at 250 °C; (**c**) wood torrefied at 350 °C; (**d**) wood torrefied at 400 °C; (**e**) wood torrefied at 400 °C with Cu(II); (**f**) wood torrefied at 400 °C with Ni(II).

DTG shows different numbers of peaks, different temperatures of combustion, and various shapes of peaks, which were found to be a difference between biochar and raw biomass. It can be concluded that structural changes appeared [14]. The highest difference between the torrefied material and raw biomass was detected at wood torrefied at 400 °C. First peaks are due to moisture loss of samples during dehydration process. The raw biomass showed huge moisture content, while the biochar content of moisture decreased with increased temperature. The mass loss of torrefied samples was higher compared to the raw wood biomass. The constant mass was reached in time intervals between 20 and 25 min. The temperature of onset of decomposition of the sample (T_i) demonstrates an increase with rising temperatures of reaction. The biochar with adsorbed nickel exhibits the highest T_i value of 399.63 °C, while the raw wood displays the lowest value of 282.52 °C. The temperature at the maximum rate of decomposition (T_b) is observed to decrease with increasing torrefaction temperature. Raw wood exhibits a T_b value of 524.34 °C, while wood torrefied at 400 °C displays a value of 440.69 °C. The presented TG and DTG curves were similar to those from the literature [15]. They found that cellulose degradation occurred first, followed by cellulose and lignin degradation.

The FTIR spectra are presented in Figure 6. The spectrum of the sample torrefied at 250 °C closely resembled that of the raw biomass, as expected given that this was the lowest torrefaction temperature. The spectra for the samples containing Cu(II) and Ni(II) were also similar; by comparing the data with literature sources [2], various functional groups in the torrefied samples were identified. Functional groups such as O–H, C–H, and N–H were detected in all samples within the wavelength range of 2920 cm^{-1} to 3350 cm^{-1}. Around 2300 cm^{-1}, C≡O, C≡N, and X=C=Y bonds (where X and Y represent C, O, N, or S) appeared in all samples except for raw wood and wood torrefied at 250 °C. This suggests that torrefaction temperature significantly influences the presence of certain functional groups. Additionally, C=C, C–O (alcohols, esters, ethers, carboxylic acids), and C–N (amines) groups were detected in all samples within the 1175 cm^{-1} to 1615 cm^{-1} range. A prominent C–O band at 1022 cm^{-1}, which is associated with cellulose and hemicellulose, showed decreasing intensity with increasing torrefaction temperature. These findings align with those of other studies [16,17].

The *HHV* results of calorimetric measurements and values of *EDR*, which were calculated according to Equation (2), and *EY*, which was calculated according to Equation (3), are presented in Table 1.

Table 1. Values of *HHV*, *EY*, and *EDR*.

Sample	*HHV* [MJ/kg]	*EY* [%]	*EDR* [/]
wood	17.581	/	/
wood torrefied at 250 °C, N_2	18.532	82.96	1.05
wood torrefied at 350 °C, N_2	23.043	47.25	1.31
wood torrefied at 400 °C, N_2	26.502	44.29	1.51
wood torrefied at 400 °C with Cu(II)	28.164	47.01	1.60
wood torrefied at 400 °C with Ni(II)	27.023	45.25	1.54

Raw wood had the lowest value among the higher heating values, which increased with increasing torrefaction temperature. Thus, temperature has a significant effect on *HHV*, and the obtained data were in accordance with the theoretical foundations of biomass torrefaction. During the torrefaction process, the content of oxygen and the proportion of volatile substances (mainly as bound water, CO_2, and CO) decreased, which contributed to an increase in the amount of carbon and consequently to an increase in the *HHV* value, hydrophobicity, grindability, and combustion properties [18]. Torrefaction improved the properties of solid fuels, leading to increased *HHVs* [19].

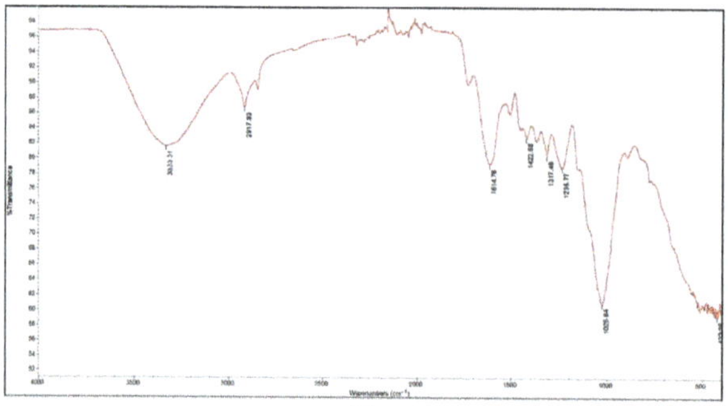

(**a**)

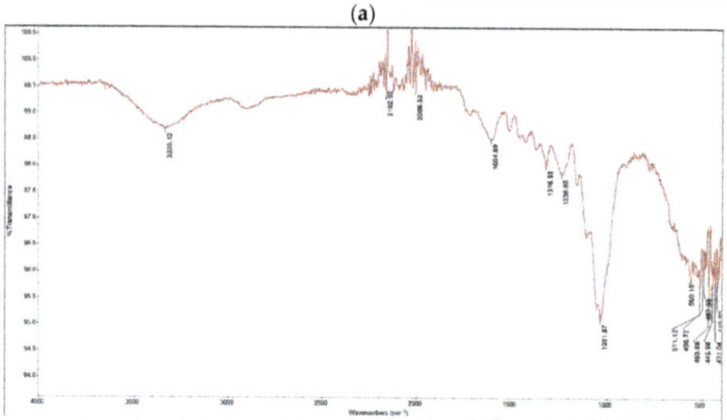

(**b**)

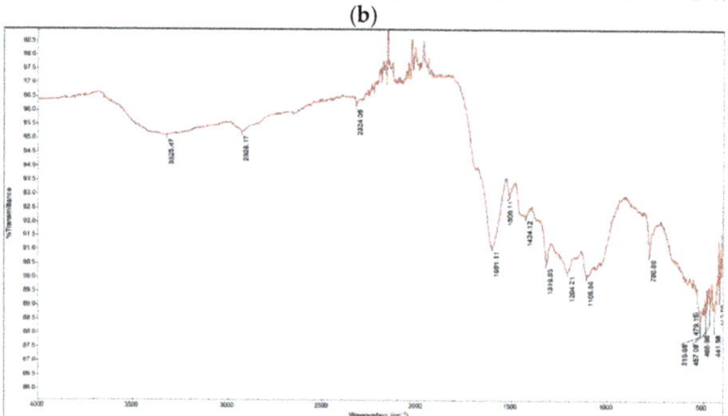

(**c**)

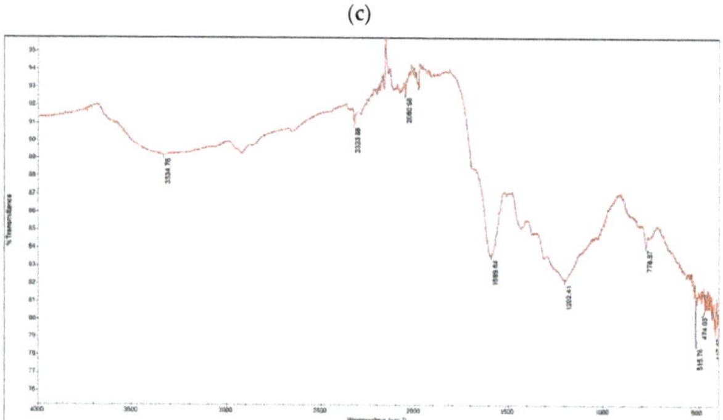

(**d**)

Figure 6. FTIR analysis results: (**a**) raw wood; (**b**) wood torrefied at 250 °C; (**c**) wood torrefied at 350 °C; (**d**) wood torrefied at 400 °C.

Lower torrefaction temperatures generally lead to higher efficiencies in terms of mass and energy yield compared with higher torrefaction temperatures. As the temperature increased from 250 to 400 °C, the mass yield decreased continuously. The highest mass and energy yields were obtained at 250 °C in CO_2, reaching 78.7% and 84.4%, respectively, which was due to the release of volatile compounds by the heat treatment. Higher temperatures tend to increase the energy density but reduce the mass and energy yield [20]. Torrefaction significantly improved the hydrophobic and grindability properties of the waste biomass; in comparison, samples treated at higher temperatures were more grindable [20]. Mass yield (*MY*) and energy yield (*EY*) represent the remaining mass in the solid phase and the energy content of torrefied biomass compared to raw biomass [21]. For all samples of torrefied biomass, the energy yield (*EY*) was greater than the mass yield (*MY*), which became even more pronounced at higher torrefaction temperatures. This phenomenon affects the increase in the heating value of torrefied biomass [5]. Increasing the energy density ratio (*EDR*) of torrefied biomass was one of the main advantages of the biomass torrefaction process [22]. The *EDR* improved with increasing process temperature, with the highest value of the energy density ratio being approximately 1.5 at 400 °C and even higher (1.6) for the same sample with adsorbed Cu(II).

3.2. The Adsorption of Moisture Results

Figure 7 shows the changes in sample mass as a function of time and relative moisture of the atmosphere. Figure 8 shows the increase in the shared mass ratio because of moisture adsorption as a function of relative humidity (*RH*). As seen from Figure 7, the percentage of mass increase was the highest in all points of relative humidity and the lowest with wood torrefied at 400 °C. The highest difference in percentage of mass was detected between the initial and final weight for raw wood. The value was determined at 5.4%. The difference between the initial and final mass for samples decreased with increased temperature. The lowest difference in the percentage of mass increase between the initial and final mass was found in wood torrefied at 400 °C, determined at 1.0%. The results are in agreement with the theoretical foundations of torrefaction, which claimed the hydrophobicity of biomass increased with the torrefaction temperature and, therefore, became more resistant to moisture adsorption [8].

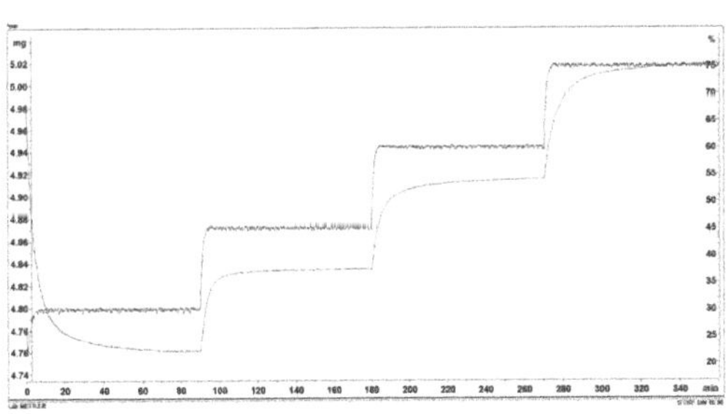

(**a**)

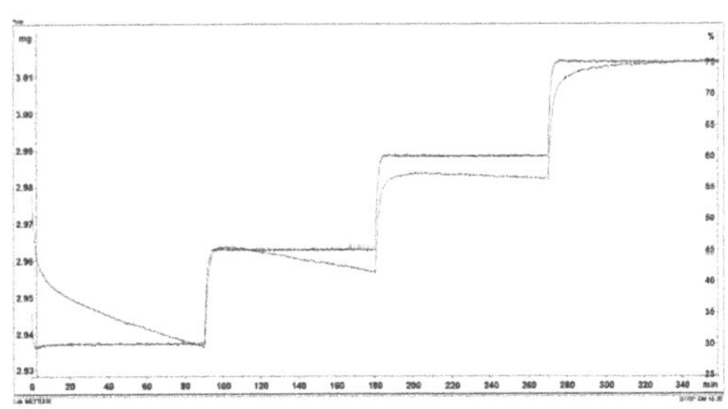

(**b**)

Figure 7. *Cont.*

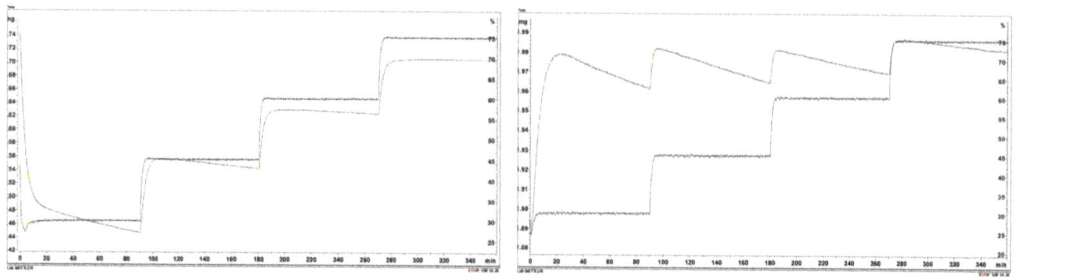

(**c**) (**d**)

Figure 7. Moisture adsorption of the following: (**a**) raw wood; (**b**) wood torrefied at 250 °C; (**c**) wood torrefied at 350 °C; (**d**) wood torrefied at 400 °C.

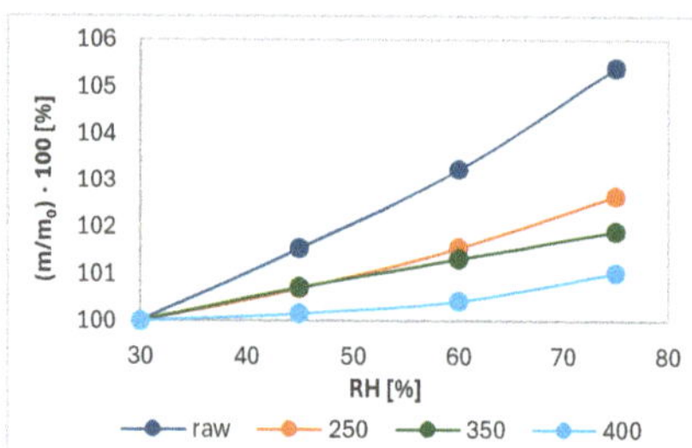

Figure 8. Percentage of increase in the mass ratio as a function of relative humidity.

3.3. The Adsorption of Cu(II) and Ni(II) Results

Figure 9 shows the variation of the concentration of the samples a) of Cu(II) and b) of Ni(II) as a function of time. The initial concentration of all solutions was 50 mg/L Cu(II) (or Ni(II)), and a decrease in concentration can be clearly seen in Figure 9a,b, indicating the successful adsorption of Cu(II) and Ni(II) onto the torrefied biomass. It was observed that Cu(II) showed a better adsorption affinity on torrefied wood than Ni(II). The adsorption for Cu(II) was higher, possibly because of strong coordination bonds formed with carboxyl groups [23]. In addition, it was observed that the sample with the highest initial mass of torrefied wood reached the highest metal adsorption onto torrefied material after 24 h of adsorption. The increment in uptake of both metals can be attributed to the constant mass transfer to the available active sites [24]. From Figure 9, it seems that the adsorption equilibrium was not reached. Therefore, the time was prolonged to 48 h. The measurement of the final concentration of both metal ions showed that the concentration did not change much compared to measurement after 24 h and sustained a plateau. Thus, it can be concluded that equilibrium was reached after 24 h.

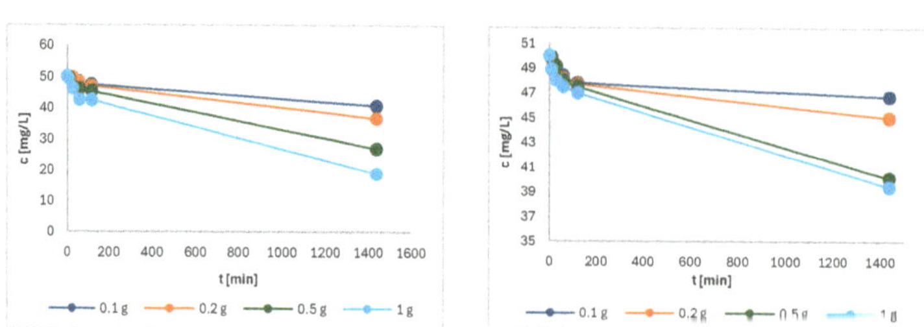

(**a**) (**b**)

Figure 9. The remaining concentration of the following in solution as a function of time: (**a**) Cu(II); (**b**) Ni(II).

Figure 10 shows the change in adsorption capacity as a function of time. The adsorption capacity of each sample increases with time, whereby the sample with the lowest mass of charcoal has the highest adsorption capacity. Figure 10 shows that samples had not yet reached equilibrium after 24 h of adsorption, suggesting that our samples require significantly more time to reach a steady state [25]. The adsorption rate is influenced by the interaction of adsorption sites on the adsorbent surface. However, subsequent measurements showed only a negligible increase in capacity after 48 h.

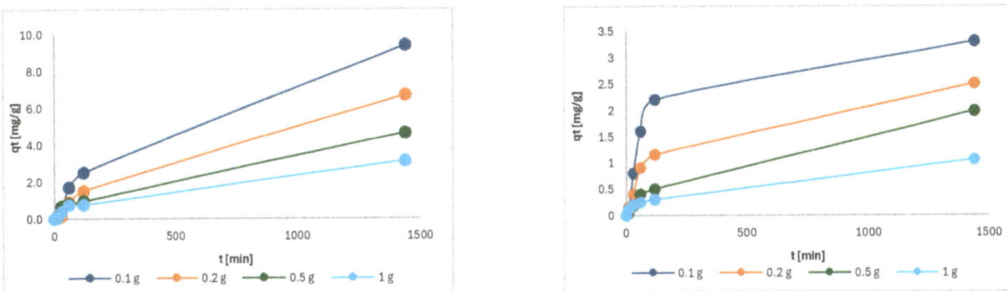

(**a**) (**b**)

Figure 10. The adsorption capacity of the following as a function of time: (**a**) Cu(II); (**b**) Ni(II).

Figure 11 shows that the pseudo-second order describes the kinetics of Cu(II) and Ni(II) adsorption well.

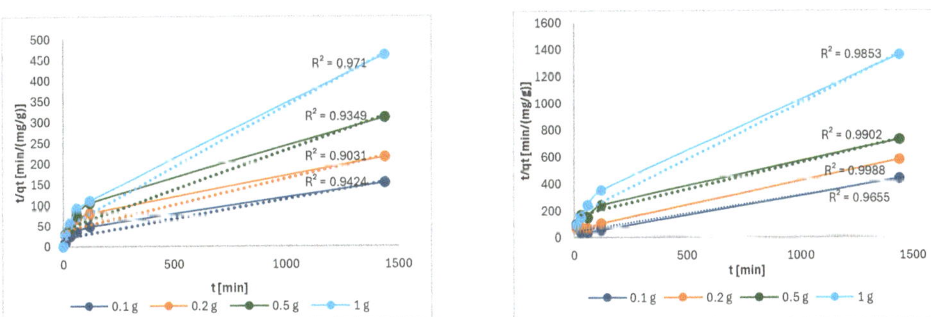

(**a**) (**b**)

Figure 11. t/qt as a function of time of the following: (**a**) Cu(II); (**b**) Ni(II).

Similar results were also obtained in other studies for Cu(II) [1] and Ni(II) [5]. A pseudo-second-order model describes chemisorption in which electron exchange takes place between the metal and the functional groups, such as the C–O band on the torrefied mixed wood [26].

Figure 12 showed the highest efficiency of Cu(II) and Ni(II) was achieved in solution where the highest mass of 1 g of torrefied material was studied. In the case of Cu(II), the efficiency was 62.4%, and for Ni(II), it was 21.2%. It was expected and is in accordance with other studies [4,5]. The efficiency was higher in Cu(II) compared to Ni(II), probably due to the smaller size of Cu(II), which may promote affinity reactions [27].

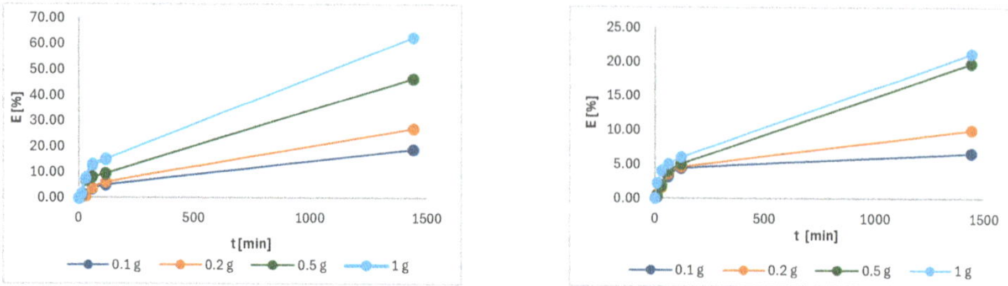

(a) (b)

Figure 12. The adsorption efficiency of the following as a function of time: (**a**) Cu(II); (**b**) Ni(II).

4. Conclusions

This study highlights the potential of torrefied biomass as an improved solid fuel with enhanced hydrophobic and energy properties. It also demonstrates its effectiveness as an adsorbent for the removal of heavy metals. The study examined the moisture adsorption behavior of torrefied biomass in comparison to raw wood, as well as the adsorption efficiency of heavy metals on torrefied material. The wood biomass was subjected to torrefaction at temperatures between 250 °C and 400 °C in an atmosphere of nitrogen. The application of lower torrefaction temperatures resulted in a notable reduction in mass, whereas the utilization of higher temperatures led to an increase in energy density. The *HHV* of the biomass exhibited an increase with rising torrefaction temperatures, thereby demonstrating enhanced energy properties.

With regard to metal adsorption, the torrefied biomass exhibited better results for Cu(II) ions, attaining an adsorption efficiency of 62.4% after 1440 min in comparison to a maximum adsorption of 21.2% for Ni(II). The kinetic analysis indicated that the adsorption process was best described by a pseudo-second-order model, which suggests that the adsorption rate is linearly related to the concentration of metal ions. The *HHV* value of torrefied material with adsorbed ions measured at 27 MJ/kg and 28 MJ/kg with Cu(II) and Ni(II), respectively, was higher compared to 26 MJ/kg of torrefied material without adsorbed heavy metal ions. This study demonstrates the potential of torrefied biomass as an improved solid fuel with enhanced hydrophobic and energetic properties, as well as its effectiveness as an adsorbent for heavy metal removal.

Author Contributions: Conceptualization, D.U. and M.S.; methodology, D.U. and M.S.; validation, A.P., M.S. and D.U.; formal analysis, I.S.; investigation, A.P.; writing—original draft preparation, I.S., D.U. and M.S.; writing—review and editing, M.S., A.P. and D.G.; supervision, A.P. and D.G. All authors have read and agreed to the published version of the manuscript.

Funding: This research was funded by the Slovenian Research Agency in the framework of Program Process Systems Engineering and Sustainable Development P2-0414 and Separation Processes and Production Design P2-0046.

Data Availability Statement: Data are contained within the article.

Conflicts of Interest: The authors declare no conflicts of interest.

References

1. Adegoke, I.A.; Ige, A.R.; Adejoba, O.R.; Aruwajoye, D.A.; James, J. Roles of Biomass in the Absorption of Heavy Metals. *Eur. J. Energy Res.* **2022**, 2, 9–13. [CrossRef]
2. Mohamed, M.A.; Jaafar, J.; Ismail, A.F.; Othman, M.H.D.; Rahman, M.A. Chapter 1-Fourier Transform Infrared (FTIR) Spectroscopy. In *Membrane Characterization*; Hilal, N., Oatley-Radcliffe, D.L., Williams, P.M., Eds.; Elsevier: Amsterdam, The Netherlands, 2017; pp. 3–29.

3. Duan, C.; Ma, T.; Wang, J.; Zhou, Y. Removal of heavy metals from aqueous solution using carbon-based adsorbents: A review. *J. Water Process Eng.* **2020**, *37*, 101339. [CrossRef]
4. Sulaiman, S.; Azis, R.S.; Ismail, I.; Man, H.C.; Yusof, K.F.M.; Abba, M.U.; Katibi, K.K. Adsorptive Removal of Copper (II) Ions from Aqueous Solution Using a Magnetite Nano-Adsorbent from Mill Scale Waste: Synthesis, Characterization, Adsorption and Kinetic Modelling Studies. *Nanoscale Res. Lett.* **2021**, *16*, 168. [CrossRef]
5. Gao, W.; He, W.; Zhang, J.; Chen, Y.; Zhang, Z.; Yang, Y.; He, Z. Effects of biochar-based materials on nickel adsorption and bioavailability in soil. *Sci. Rep.* **2023**, *13*, 5880. [CrossRef] [PubMed]
6. Hyk, W.; Święcicka, D.; Garboś, S. Application of mixed (bimodal) distribution to human health risk assessment of Cu and Ni in drinking water collected by RDT sampling method from a large water supply zone. *Microchem. J.* **2013**, *110*, 465–472. [CrossRef]
7. Zhou, Q.; Shen, Y.; Gu, X. Progress in torrefaction pretreatment for biomass gasification. *Green Chem.* **2024**, *26*, 9652–9670. [CrossRef]
8. Grycova, B.; Pryszcz, A.; Krzack, S.; Klinger, M.; Lestinsky, P. Torrefaction of biomass pellets using the thermogravimetric analyser. *Biomass Convers. Biorefinery* **2021**, *11*, 2837–2842. [CrossRef]
9. Ivanovski, M.; Goricanec, D.; Krope, J.; Urbancl, D. Torrefaction pretreatment of lignocellulosic biomass for sustainable solid biofuel production. *Energy* **2022**, *240*, 122483. [CrossRef]
10. van der Stelt, M.J.C.; Gerhauser, H.; Kiel, J.H.A.; Ptasinski, K.J. Biomass upgrading by torrefaction for the production of biofuels: A review. *Biomass Bioenergy* **2011**, *35*, 3748–3762. [CrossRef]
11. Odusote, J.K.; Adeleke, A.A.; Lasode, O.A.; Malathi, M.; Paswan, D. Thermal and compositional properties of treated Tectona grandis. *Biomass Convers. Biorefinery* **2019**, *9*, 511–519. [CrossRef]
12. Petrovič, A.; Simonič, M. Removal of heavy metal ions from drinking water by alginate-immobilised Chlorella sorokiniana. *Int. J. Environ. Sci. Technol.* **2016**, *13*, 1761–1780. [CrossRef]
13. Wei, M.; Chen, J.; Wang, X. Removal of arsenic and cadmium with sequential soil washing techniques using Na2EDTA, oxalic and phosphoric acid: Optimization conditions, removal effectiveness and ecological risks. *Chemosphere* **2016**, *156*, 252–261. [CrossRef] [PubMed]
14. Li, S.; Chen, G. Thermogravimetric, thermochemical, and infrared spectral characterization of feedstocks and biochar derived at different pyrolysis temperatures. *Waste Manag.* **2018**, *78*, 198–207. [CrossRef]
15. Chen, D.; Gao, A.; Cen, K.; Zhang, J.; Cao, X.; Ma, Z. Investigation of biomass torrefaction based on three major components: Hemicellulose, cellulose, and lignin. *Energy Convers. Manag.* **2018**, *169*, 228–237. [CrossRef]
16. Oyebode, W.A.; Ogunsuyi, H.O. Impact of torrefaction process temperature on the energy content and chemical composition of stool tree (Alstonia congenisis Engl) woody biomass. *Curr. Res. Green Sustain. Chem.* **2021**, *4*, 100115. [CrossRef]
17. So, C.-L.; Eberhardt, T.L. FTIR-based models for assessment of mass yield and biofuel properties of torrefied wood. *Wood Sci. Technol.* **2018**, *52*, 209–227. [CrossRef]
18. Dyjakon, A.; Noszczyk, T.; Sobol, Ł.; Misiakiewicz, D. Influence of Torrefaction Temperature and Climatic Chamber Operation Time on Hydrophobic Properties of Agri-Food Biomass Investigated Using the EMC Method. *Energies* **2021**, *14*, 5299. [CrossRef]
19. Matali, S.; Rahman, N.; Idris, S.; Yaacob, N.; Alias, A. Lignocellulosic Biomass Solid Fuel Properties Enhancement via Torrefaction. *Procedia Eng.* **2016**, *148*, 671–678. [CrossRef]
20. Ivanovski, M.; Urbancl, D.; Petrovič, A.; Stergar, J.; Goričanec, D.; Simonič, M. Improving Lignocellulosic and Non-Lignocellulosic Biomass Characteristics through Torrefaction Process. *Appl. Sci.* **2022**, *12*, 12210. [CrossRef]
21. Šantl, N.; Stergar, J.; Bozicko, M.; Goričanec, D.; Urbancl, D.; Petrovič, A. The utilisation of thermally treated poultry farm waste for energy recovery and soil application. *Renew. Energy* **2024**, *221*, 119809. [CrossRef]
22. Świechowski, K.; Liszewski, M.; Bąbelewski, P.; Koziel, J.A.; Białowiec, A. Oxytree Pruned Biomass Torrefaction: Mathematical Models of the Influence of Temperature and Residence Time on Fuel Properties Improvement. *Materials* **2019**, *12*, 2228. [CrossRef] [PubMed]
23. Premchand, P.; Mead, S.; Fino, D.; Demichelis, F.; Bensaid, S.; Chiaramonti, D.; Antunes, E. Sustainable valorisation of cigarette butts waste through pyrolysis: An insight into the pyrolytic products and subsequent aqueous heavy metals removal by pyrolytic char. *Chem. Eng. Sci.* **2025**, *302*, 120906. [CrossRef]
24. Choudhary, M.; Kumar, R.; Neogi, S. Activated biochar derived from Opuntia ficus-indica for the efficient adsorption of malachite green dye, Cu+2 and Ni+2 from water. *J. Hazard. Mater.* **2020**, *392*, 122441. [CrossRef]
25. Karagoz, S.; Tay, T.; Ucar, S.; Erdem, M. Activated carbons from waste biomass by sulfuric acid activation and their use on methylene blue adsorption. *Bioresour. Technol.* **2008**, *99*, 6214–6222. [CrossRef] [PubMed]

26. Kończyk, J.; Kluziak, K.; Kołodyńska, D. Adsorption of vanadium (V) ions from the aqueous solutions on different biomass-derived biochars. *J. Environ. Manag.* **2022**, *313*, 114958. [CrossRef]
27. Zhou, Z.; Xu, Z.; Feng, Q.; Yao, D.; Yu, J.; Wang, D.; Lv, S.; Liu, Y.; Zhou, N.; Zhong, M.-E. Effect of pyrolysis condition on the adsorption mechanism of lead, cadmium and copper on tobacco stem biochar. *J. Clean. Prod.* **2018**, *187*, 996–1005. [CrossRef]

metals

MDPI

Article

Rare Earth Metal Ion-Associates in Ln^{3+}—CO_3^{2-}—H_2O System

Tatiana Litvinova [1], Stepan Gerasev [2,*], Vasiliy Sergeev [1] and Egor Lidanovskiy [3]

1 The Department of General and Physical Chemistry, Empress Catherine II Saint Petersburg Mining University, 199106 Saint Petersburg, Russia; litvinova_te@pers.spmi.ru (T.L.); sergev_vv2@pers.spmi.ru (V.S.)
2 The Metallurgy Department, Empress Catherine II Saint Petersburg Mining University, 199106 Saint Petersburg, Russia
3 The Department of Chemical Engineering and Energy Processing, Empress Catherine II Saint Petersburg Mining University, 199106 Saint Petersburg, Russia; s220186@stud.spmi.ru
* Correspondence: s235035@stud.spmi.ru

Abstract: This study focused on the nature of rare earth metal complex compounds that can form during the carbonate–alkaline processing of industrial waste materials, such as phosphogypsum and red mud, at 70–100 °C and 1–10 atm. Experimental findings revealed that the dissolution of synthetic carbonates of rare earth elements (REEs) in a concentrated carbonate-ion medium (3 mol/L) leads to the formation of ion-associates of varying strengths. Light (lanthanum, praseodymium, and neodymium) and medium (samarium) REE groups exhibited a tendency to form loose ion-associates, whereas heavy REEs (terbium, dysprosium, holmium, erbium, thulium, lutetium, and yttrium) formed close ion-associates. To confirm the existence of these ion-associates, the specific conductivity of solutions was measured after dissolving thulium (III) and samarium (III) carbonates at phase ratios ranging from 1:2000 g/mL to 1:40 g/mL in a potassium carbonate medium. The decay of ion-associates, leading to the precipitation of rare earth metal (III) carbonates, was tested in an ammonium carbonate medium. Thermal decomposition of ammonium carbonate at 70–75 °C during 1–4 h was accompanied by full rare earth carbonates' sedimentation and its in-the-way separation into groups because of the varied strength of ion-associates. The results of this study provide a basis for developing processes to separate rare earth metals into groups during their carbonate–alkaline extraction into solution.

Keywords: rare earth metals; lanthanides; ion-associates; carbonate complexes; rare earth carbonates; solubility; ammonium carbonate; potassium carbonate

1. Introduction

Rare earth metals (REMs) are among the key types of strategic mineral resources, comprising 17 elements: 15 lanthanides, yttrium, and scandium [1–3]. The unique physicochemical properties of REMs enable their use in the production of permanent magnets, batteries, lasers, construction materials, alloys, catalysts, optical glasses, and various other products and intermediates. These materials are particularly essential in military industries, mechanical engineering, and medicine [4–6].

Among the existing classifications of REMs, the following are the most widely applied: [7,8]:

1. By Mass
   - Light (La–Nd);
   - Medium (Sm–Gd);
   - Heavy (Dy–Lu, Y).

2. By Importance, to eliminate the disbalance between global production and consumption of individual REEs
    - Critical (Nd, Eu, Tb, Dy, Er, Y);
    - Non-critical (La, Pr, Sm, Gd, Lu);
    - Surplus (Ce, Ho, Tm, Yb).

Monazite, xenotime, bastnasite, loparite, and other minerals serve as natural sources of surplus, non-critical, and critical REMs alike [9–11]. However, the generation of significant amounts of REM-containing industrial waste, such as phosphogypsum (PG) and red mud (RM), has made the recycling of secondary raw materials increasingly important [12–14]. Globally, PG reserves are estimated at 7–8 billion tons, while RM exceeds 4 billion tons, with REM content in these materials up to 0.9% [15–17].

Although the extraction of REMs and their compounds from traditional sources is predominantly achieved through acid-based methods, the processes developed over the past decades for processing PG and RM using mineral acids lack sufficient economic and/or environmental efficiency [18–20]. In this regard, the alternative carbonate–alkaline method—treatment with ammonium carbonate or alkali metal carbonate solutions—best meets modern requirements. However, it still requires optimization in terms of REMs' extraction into solution and their subsequent separation [21,22].

The development of technical solutions for REM recovery remains constrained by the lack of definitive data on the forms of lanthanides, scandium, and yttrium in a carbonate-ion medium. Exceptions include a limited number of studies, such as those by Millero [23], Byrne [24], and others.

It is known that REMs in PG and RM can exist in the form of various sparingly soluble compounds: fluorides, sulfates, and phosphates in PG, and hydroxides and carbonates in RM [25,26]. REM cations (Ln^{3+}), as well as the corresponding precipitates, are capable of forming carbonate complexes of variable composition, $Ln(CO_3)_n{}^{3-2n}$, under the influence of an excess of carbonate ions. However, the structures of the resulting compounds remains uncertain [27–29].

On the one hand, the interaction of Ln^{3+} with $CO_3{}^{2-}$ (1) may involve the formation of classical coordination compounds, consisting of a central Ln ion and $CO_3{}^{2-}$ ligands. Under standard conditions, for $n = 1$, the stability constant K_1 ranges from 6.52 to 9.15, and for $n = 2$ K_2 it ranges from 11.58 to 17.41 [30–32]. Stability constants are presented in Table 1. On the other hand, the presence of a stability constant alone does not provide definitive evidence of a coordination compound.

Table 1. Stability constants of REM carbonate complexes [32].

REM	Ohta and Kawabe		Liu and Burne		REM	Ohta and Kawabe		Liu and Burne	
	$\lg K_1$	$\lg K_2$	$\lg K_1$	$\lg K_2$		$\lg K_1$	$\lg K_2$	$\lg K_1$	$\lg K_2$
La	8.33	12.52	6.52	11.58	Dy	8.95	14.38	7.33	13.18
Ce	8.58	13.06	6.86	12.05	Ho	8.96	14.48	7.32	13.27
Pr	8.73	13.43	7.03	12.37	Er	9.02	14.62	7.38	13.39
Nd	8.75	13.59	7.08	12.46	Tm	9.10	14.79	7.45	13.54
Sm	8.90	13.95	7.25	12.82	Yb	9.15	14.94	7.58	13.56
Eu	8.86	14.02	7.26	12.91	Lu	9.13	14.96	7.53	13.64
Gd	8.78	13.95	7.17	12.76	Y	8.88	14.29	7.25	12.90
Tb	8.88	14.22	7.23	13.05	Sc	-	17.41		-

Moreover, the nature of the bond between a REM ion and a carbonate anion remains unclear. Some researchers [33,34] suggest the possible formation of a complex in the form of an ion-associate (Figure 1), as follows:

$$Ln^{3+}_{(aq)} + nCO^{2-}_{3(aq)} \rightarrow Ln(CO_3)^{3-2n}_{n(aq)} \quad (1)$$

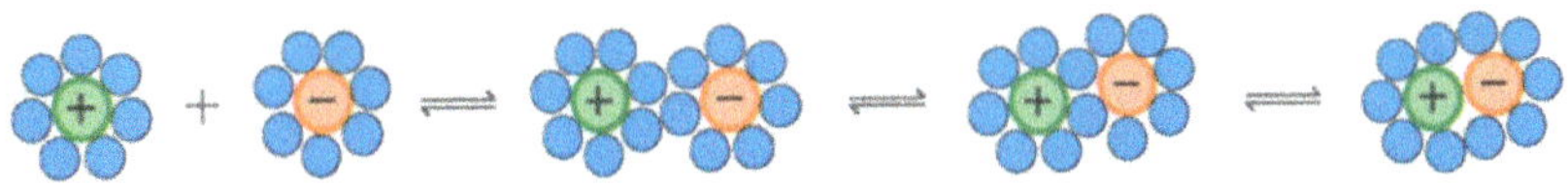

Figure 1. Mechanism of contact (close) ion pair formation [33].

These forms are composed of particles held together by electrostatic attraction forces and comply with Coulomb's law. Such behavior of lanthanides has been confirmed, for instance, for Yb^{3+} and Lu^{3+} ions in a Cl^- medium at concentrations exceeding 2 mol/L [34–36].

Thermodynamic evidence for the formation of ion pairs, triplets, and other structures lies in the high positive entropy values associated with the formation of ion-associates [33]. Due to the limited availability of literature data on the thermodynamics of the association processes involving REM cations and carbonate anions, the most analogous association process is (2), where the entropy of associate formation under standard conditions is 83 ± 12 J/(mol·K) [37]:

$$CaCO_{3(s)} + CO^{2-}_{3(aq)} \rightarrow Ca(CO_3)^{2-}_{2(aq)} \quad (2)$$

The entropy of formation for REM carbonate complexes under standard conditions, calculated using the Latimer–Powell equation, ranges from 159.9 J/(mol·K) to 266.9 J/(mol·K), which theoretically confirms the formation of $Ln(CO_3)_n^{3-2n}$ as an ion-associate [38].

According to the literature, the formation of ion-associates in solution does not lead to changes in its conductivity. Ion pairing or association occurs when oppositely charged ions are closer than a certain critical distance. They then can act as a neutral form and hence do not contribute to conductivity [34,39].

Based on the above, determining the type of soluble REM form in a carbonate-ion medium facilitates the development of approaches for their subsequent extraction and separation from the productive solution. Therefore, the aim of this study was to establish the feasibility of REM ion-associates existing in the Ln^{3+}—$CO_3{}^{2-}$—H_2O system.

2. Materials and Methods

Measuring the specific electrical conductivity (SEC) of the system formed during the dissolution of synthetic REM (III) carbonates in a potassium carbonate medium facilitated confirming or refuting the presence of associates.

The use of $Ln_2(CO_3)_3$ as the material for conductometric studies eliminated the influence of foreign anions and/or added water on SEC values. The selected concentration of the potassium carbonate solution, 3 mol/L, corresponded to the range in which SEC increases with concentration (Figure 2) and remained within the sensitivity limits of the device («Anion 4100»). The main technical characteristics of «Anion 4100» are its measurement range of SEC from 10^{-6} S/cm to 10^{-1} S/cm and that its limit of permissible basic relative SEC measurement error of solutions is no more than 2%.

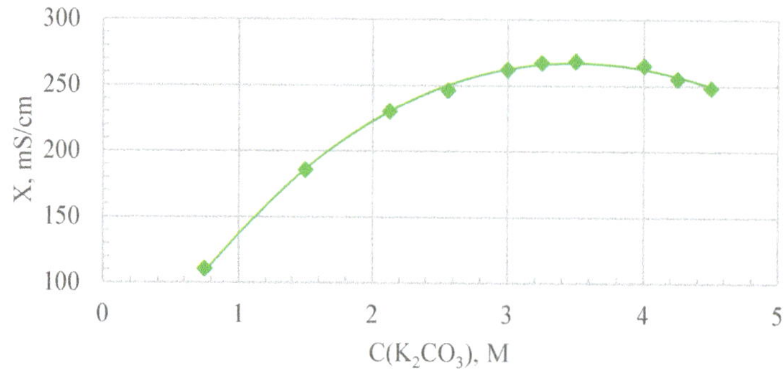

Figure 2. Dependence of solution-specific electrical conductivity on K_2CO_3 concentration at 25 °C.

This study of the effect of potassium, sodium, and ammonium cations on REM carbonate solubility in 0.5–3 M of carbonate ion medium was undertaken using cerium carbonate as an example under standard conditions and a liquid-to-solid ratio of 2000 mL/g (Figure 3).

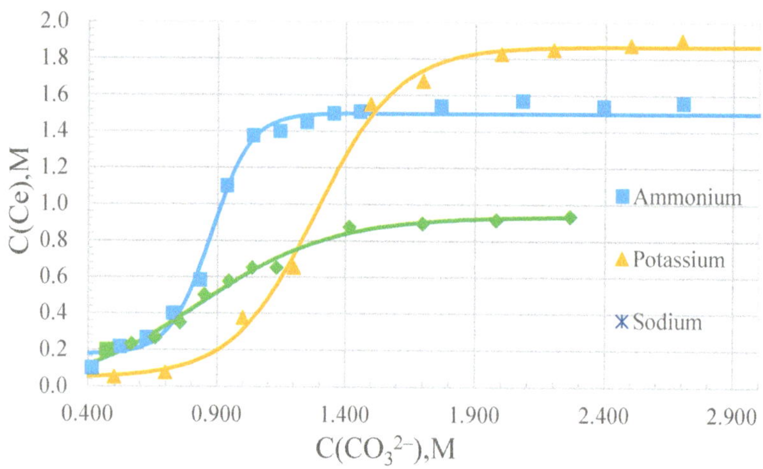

Figure 3. Influence of carbonate ion concentration on cerium (III) carbonate solubility at 25 °C.

The main laboratory equipment for all declared experiments was a HEL Automate II Parallel Reactor System (H.E.L. Group, London, UK) fitted with temperature and stirring speed sensors. Thus, the mixtures' mixing was organized thermostatically.

This series of experiments was conducted under standard conditions by mixing solid REM (III) carbonate with a potassium carbonate solution at solid-to-liquid ratios from 1:2000 g/mL to 1:40 g/mL. Each experiment was carried out for 12 h with stirring at 200 rpm to reach an equilibrium characterized by the complete dissolution of the precipitate.

Another potential method for identifying the soluble forms of REMs is monitoring the reduction in carbonate ion concentration after the complexation process. When a complex ion, $Ln(CO_3)_n^{3-2n}$, is formed via covalent bonds, a decrease in CO_3^{2-} concentration does not lead to its decay due to the relatively high stability constants. In contrast, the presence of ion-associates would result in their decomposition, as ion pairs and other associates are prone to it [40]. As among soluble carbonates only ammonium carbonate tends to decompose at temperatures below the boiling point of the solution (60–70 °C), the reduction in concentration due to thermal decomposition was studied in this medium. In this series of experiments, solutions of REM nitrate (0.4 mol/L) and ammonium carbonate (3 M) were mixed during 5–10 min at a volume ratio of 1:20 mL/mL. After precipitation and subsequent dissolution, the solution was heated to 70–75 °C. The secondary precipitate formed during ammonium carbonate decomposition was quantified using a gravimetric method, while the REM concentration in the final solution was determined via complexometric titration with Trilon B in the presence of Arsenazo (III) using Equation (3):

$$C_{Ln} = \frac{C_{TB} \times V_{TB}}{V_{Ln}} \tag{3}$$

where C_{Ln}—normal concentration of REM (III), eq/L; V_{Ln}—volume of the aliquot, L; C_{TB}—normal concentration of Trilon B (III), eq/L; and V_{TB}—volume of the titrant, L.

Among the possible insoluble REM (III) compounds that could form upon the decay of the ion-associate, only $Ln_2(CO_3)_3$ was considered. For precipitates obtained, identification was determined via SEM-EDX analysis by scanning electron microscope VEGA 3 LMH with an energy dispersive X-ray microanalysis system (TESCAN Essence, Brno, Czech Republic). Thus, the REM carbonate yield can be calculated using Equation (4):

$$\eta_{Ln_2(CO_3)_3} = \frac{m^p_{Ln_2(CO_3)_3}}{m^t_{Ln_2(CO_3)_3}} \times 100\% = \frac{m^p_{Ln_2(CO_3)_3} \times 2}{C_{Ln(NO_3)_3} \times V_{Ln(NO_3)_3} \times M_{Ln_2(CO_3)_3}} \times 100\% \tag{4}$$

where $\eta_{Ln_2(CO_3)_3}$—yield of $Ln_2(CO_3)_3$, %; $m^p_{Ln_2(CO_3)_3}$—the mass of $Ln_2(CO_3)_3$ after drying, g; $m^t_{Ln_2(CO_3)_3}$—stoichiometrically calculated mass of $Ln_2(CO_3)_3$, g; $C_{Ln(NO_3)_3}$—the concentration of the $Ln(NO_3)_3$, solution, mol/L; $V_{Ln(NO_3)_3}$—the volume of the $Ln(NO_3)_3$, solution, L; and $M_{Ln_2(CO_3)_3}$—the molar mass of $Ln_2(CO_3)_3$.

The concentration of ammonium carbonate in the final solution was determined using conductometric titration with sodium hydroxide.

A short thermodynamic assessment of performed processes was completed. The Gibbs energy change during these reactions to REM carbonates obtained through its ion-associates was estimated by applying the Hess law. A database of the Chemistry Faculty of Moscow State University «https://www.chem.msu.ru/cgi-bin/tkv.pl» (accessed on 17 February 2025) was chiefly used for this purpose.

3. Results

The experimentally determined SEC of the aqueous potassium carbonate solution at 25 °C was 263.2 mS/cm. Upon adding a sample of thulium (III) carbonate to this solution at solid-to-liquid ratios of 1:2000 g/mL, 1:100 g/mL, and 1:40 g/mL, the electrical conductivity of the system decreased, reaching a minimum 2 min after mixing. The conductivity values were 259.1 mS/cm, 253.1 mS/cm, and 252.2 mS/cm, respectively. Whereafter, they gradually returned to the initial conductivity value of 263.2 mS/cm by 4, 6, and 9 min (Figure 4).

The yields of REM (III) carbonates were calculated after heating the solutions that contained the rare earth metals (III) ion-associates for 60, 120, and 240 min (Figure 5), as follows:

1. Yield of $Ln_2(CO_3)_3$ after 60 min: La—98.8 %; Nd—98.3 %; Pr—99.6 %; Sm—97.9 %; Tb—99.4 %; Dy—87.2 %; Ho—63.3 %; Y (Er)—51.6 %; Tm—0 %; Lu—0 %.
2. Yield of $Ln_2(CO_3)_3$ after 120 min: La—98.8 %; Nd—98.3 %; Pr—99.6 %; Sm—97.9 %; Tb—99.4 %; Dy—98.4 %; Ho—74.5 %; Y (Er)—60.9 %; Tm—5 %; Lu—5 %.
3. Yield of $Ln_2(CO_3)_3$ at 240 min reached approximately 100 %, including Tm and Lu.

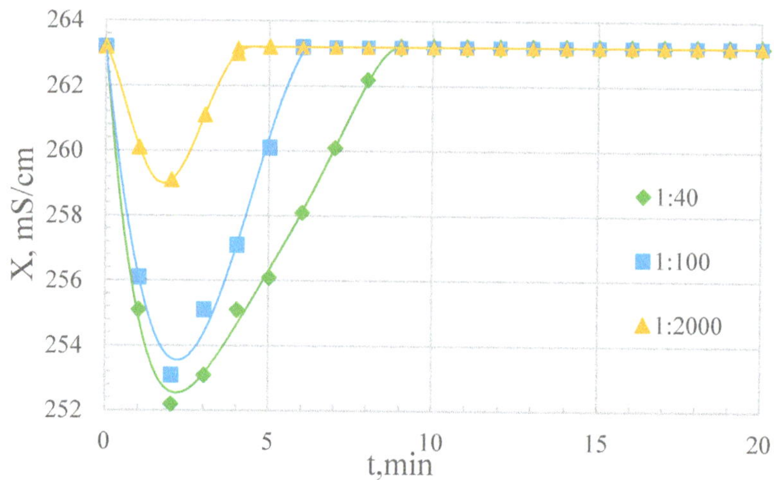

Figure 4. Change in specific electrical conductivity upon adding $Tm_2(CO_3)_3$ to K_2CO_3 solution.

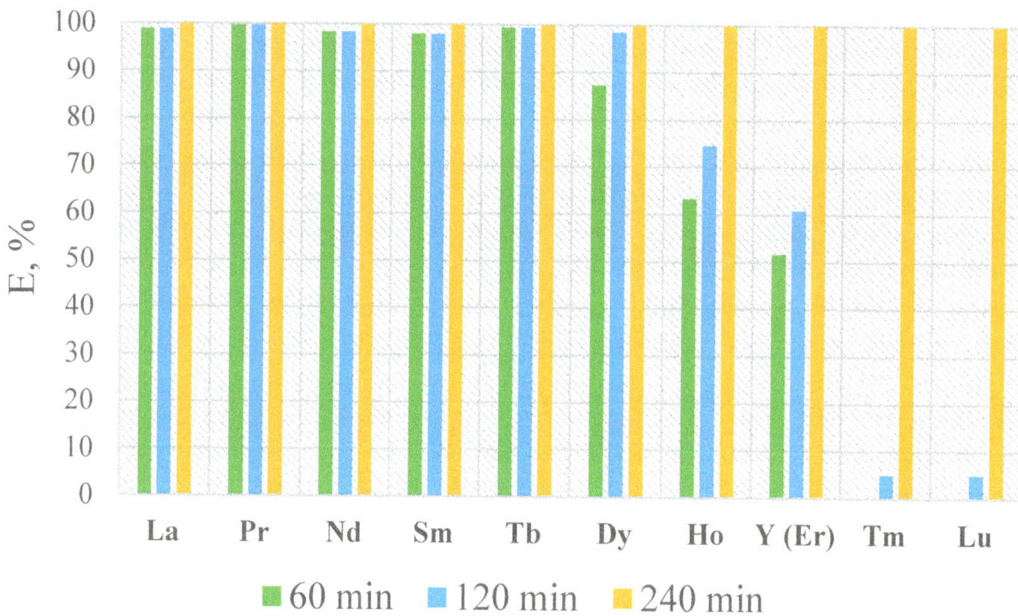

Figure 5. Yield of $Ln_2(CO_3)_3$ upon decay of ion-associates.

The kinetics of ammonium carbonate decomposition were studied using a model solution (initial concentration 3 mol/L) at a temperature of 70–75 °C over a period of 3 h. Based on data shown in Figure 6, the concentration of $(NH_4)_2CO_3$ decreased from 3 mol/L to 1.9 mol/L after 1 h, and from 3 mol/L to 1.05 mol/L after 2 h.

The results of SEM-EDX analysis of precipitates produced in the process of REM ion-associates decay are illustrated in Figures 7 and 8 using thulium precipitate as an example. The thulium precipitate was composed of 71.48 wt. % thulium (Tm), 23.63 wt. % oxygen (O), and 4.89 wt. % carbon (C).

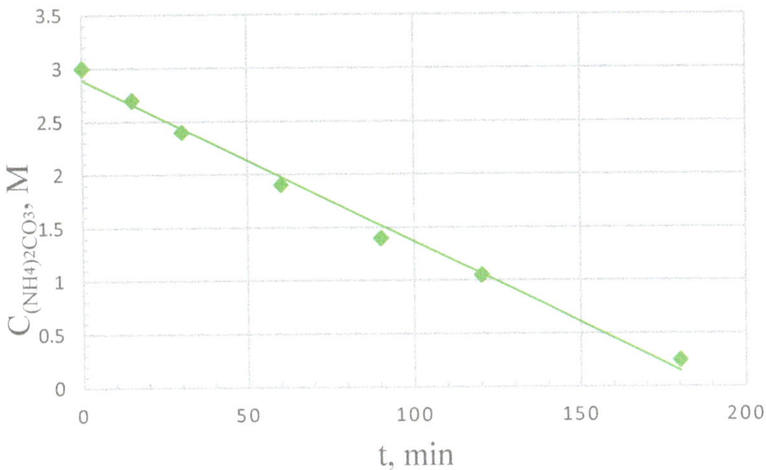

Figure 6. Kinetics of $(NH_4)_2CO_3$ decomposition.

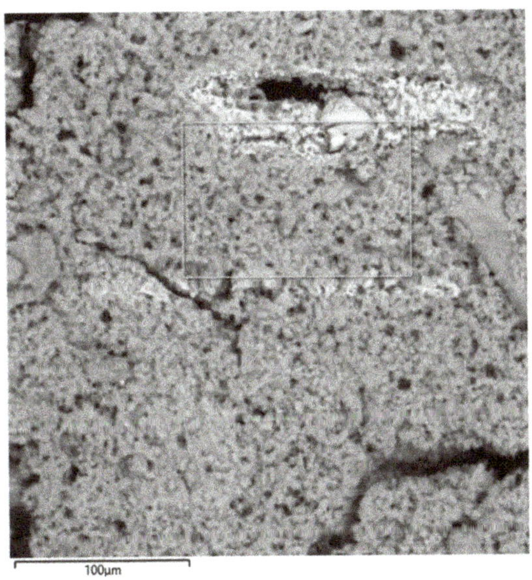

Figure 7. SEM image of thulium precipitate during its ion-associate decay.

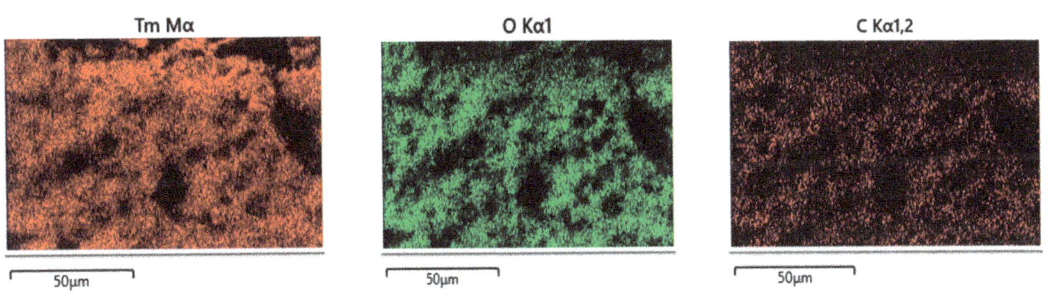

Figure 8. SEM-EDX mapping of thulium precipitate during its ion-associate decay.

Preliminary experiments in the field of REM leaching from technogenic phosphogypsum were carried out. Leaching was organized by a 4 M carbonate solution that was stirred for 3 h at a temperature of 90 °C and an L:S 100 mL per 1 g of PG. The degrees of extraction were as follows: La—40.1%, Ce—58.2%, Nd—62.5%, Dy—64.0%, Er—66.4%, Y—72.9%, and Yb—76.8%.

4. Discussion

The difference in cerium (III) carbonate solubility in potassium, sodium, and ammonium carbonate medium indicates the influence of the cation nature, which is characteristic of the association process.

The initial change in electrical conductivity, as shown in Figure 4, was caused by the formation of a suspension, with a higher amount of added powder (such as 0.1 g, 2 g, and 5 g) leading to a greater drop in the conductivity of the solution. After 2 min, an increase in specific electrical conductivity was observed, accompanied by the dissolution of $Tm_2(CO_3)_3$. Complete dissolution of the precipitate, depending on the solid-to-liquid ratio, was observed within 5–10 min. Complete dissolution of the precipitate was characterized by the return to the initial SEC value (263.2 mS/cm) for all considered phase ratios, confirming the formation of a thulium carbonate complex in the form of an ion-associate.

The behavior of carbonate complexes of light REEs (lanthanum and neodymium), medium-mass REEs (samarium and gadolinium), and heavy REEs (terbium, dysprosium, holmium, thulium, and yttrium) varied with a decrease in carbonate ion concentration. When heating solutions containing La (III), Nd (III), and Sm (III) at 70–75 °C, turbidity was observed at 15 min into the decomposition of ammonium carbonate, accompanied by the formation of the corresponding REE (III) carbonates. For Tb (III) and Dy (III), precipitation began at 20 min, for Ho (III) it began at 35 min, for Y at 40 min, for Tm at 110 min, and for Lu at 120 min.

According to the SEM-EDX analysis results, there was a possibility of the REM carbonate formation, or a like compound, in the ion-associate decay process. In a thulium carbonate, a mass fraction must be 65.24 wt. %, but it was in fact 71.48 wt. %. Therefore, the excursion was in the area of 6 wt. %

The tendency of REM carbonate complexes to decompose, forming a secondary precipitate such as $Ln_2(CO_3)_3$ upon a decrease in CO_3^{2-} concentration, serves as additional evidence that the soluble forms of Ln(III) in concentrated solutions of $(NH_4)_2CO_3$, K_2CO_3, or Na_2CO_3 are ion-associates.

Carbonate ion concentration decreases led to a chemical equilibrium shift towards the formation of REM carbonates. Moreover, the difference in the time required for the disintegration of ion-associates can be explained by the difference in individual elements' stability constants; i.e., a higher value of the stability constant required longer heating for a significant decrease in the carbonate ion in the solution.

The variation in the time required for the decay of ion-associates was due to differences in the strength of electrostatic attraction between anions and individual groups of REMs (III). Ions with a higher charge and smaller size exhibit stronger electrostatic interactions [40]. Therefore, the shorter decay times of light and medium REM associates resulted from their lower stability, indirectly suggesting a predominance of loose associates. In contrast, the strong associates formed by heavy REMs were primarily characterized as close (contact) associates.

The concentration of $(NH_4)_2CO_3$ in the productive solution after the decay of ion-associates was assumed to be according to Figure 6. An example of the material balance is presented in Table 2, characterizing the decay process of the terbium (III) ion-associate during heating for 1 h.

Table 2. Material balance of process for terbium carbonate formation.

Input			Output		
№	Item	Mass, g	№	Item	Mass, g
1	Ammonium carbonate solution:	219.40	1	Productive solution:	199.00
	Ammonium carbonate	57.60		Ammonium carbonate	34.10
	Water	161.80		Ammonium nitrate	0.96
2	Terbium nitrate (III) solution:	11.18		Water	163.94
	Terbium nitrate (III)	1.38	2	Water	11.06
	Water	9.8		Water (evaporation)	6.55
	Total:	230.58		Water from Equation (5)	4.51
			3	Ammonia	8.51
			4	Carbon dioxide	11.02
			5	Terbium carbonate (III)	0.99
				Total:	230.58

The material balance accounts for the processes described by Equations (8) and (12), which are end-to-end, and consist of Equations (5)–(11).

$$\text{Precipitation}: \ 2\text{Ln}^{3+}_{(aq)} + 3\text{CO}^{2-}_{3(aq)} \rightarrow \text{Ln}_2(\text{CO}_3)_{3(s)} \tag{5}$$

$$\text{Ion-association (dissolution)}: \ \text{Ln}_2(\text{CO}_3)_{3(s)} + \text{CO}^{2-}_{3(aq)} \rightarrow 2\text{Ln}(\text{CO}_3)^{-}_{2(aq)} \tag{6}$$

$$\text{Decay of ion-associates}: \ 2\text{Ln}(\text{CO}_3)^{-}_{2(aq)} \rightarrow \text{Ln}_2(\text{CO}_3)_{3(s)} + \text{CO}^{2-}_{3(aq)} \tag{7}$$

$$\text{End-to-end equation}: \ 2\text{Ln}^{3+}_{(aq)} + 3\text{CO}^{2-}_{3(aq)} \rightarrow \text{Ln}_2(\text{CO}_3)_{3(s)} \tag{8}$$

$$\text{Hydrolysis}: \ 2\text{NH}^{+}_{4(aq)} + \text{CO}^{2-}_{3(aq)} + 2\text{H}_2\text{O}_{(l)} \rightarrow 2\text{NH}_3{\cdot}\text{H}_2\text{O}_{(l)} + \text{H}_2\text{CO}_{3(l)} \tag{9}$$

$$\text{Decomposition}: \ \text{H}_2\text{CO}_{3(l)} \rightarrow \text{CO}_{2(g)} + \text{H}_2\text{O}_{(l)} \tag{10}$$

$$\text{Decomposition}: \ \text{NH}_3{\cdot}\text{H}_2\text{O}_{(l)} \rightarrow \text{NH}_{3(g)} + \text{H}_2\text{O}_{(l)} \tag{11}$$

$$\text{End-to-end equation}: \ 2\text{NH}^{+}_{4(aq)} + \text{CO}^{2-}_{3(aq)} \rightarrow 2\text{NH}_{3(g)} + \text{CO}_{2(g)} + \text{H}_2\text{O}_{(l)} \tag{12}$$

Figure 9 shows the calculated values of Gibbs energy for mono- and di-carbonate REM complexes in carbonate–carbonate systems. Similar values were obtained when REM sulfates and phosphates were leached by carbonate solution [41]. Therefore, complexation processes are possible.

If Ln is Tb, the Gibbs energy change at standard conditions during these processes as ΔG° (5) or (8) = −2.27 kJ/mol, ΔG° (6) ≈ −160 kJ/mol, and ΔG° (7) = 149.46 kJ/mol. The reactions from (9) to (12) are characterized as follows: ΔG° (9) = 230.70 kJ/mol, ΔG° (10) = −104.02 kJ/mol, ΔG° (11) = 0.28 kJ/mol, and ΔG° (12) = 127.24 kJ/mol.

On one hand, a positive value of ion-associate decay (7) reflects a theoretical stability of the complex compound. However, on the other hand, a carbonate ion decrease and its subsequent removal from the system lead to an equilibrium shift towards the REM carbonates' formation. This can be explained either by the formation of intermediate compounds or by complexation as an ion-association without strong covalent bonds.

Based on the analysis of the temperature effect at 60–75 °C on the decay of ion-associates (due to the decrease in ammonium carbonate concentration), the formation of rare earth metal (III) carbonates up to terbium occurred in a shorter time (τ_2) and with a higher yield compared to heavier rare earth metals, whose ion-associates decayed over a longer time (τ_1). Thus, the mechanism for the formation of rare earth metal (III) carbonates can be characterized by the scheme shown in Figure 10.

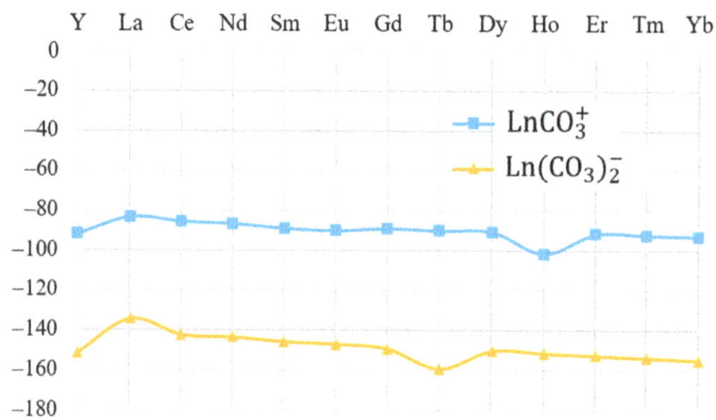

Figure 9. Gibbs energy changes during rare earth metal (III) carbonates' dissolution in carbonate-ion medium.

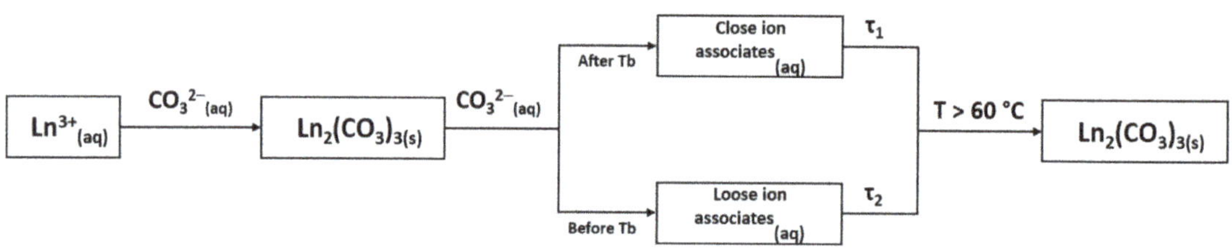

Figure 10. Mechanism of formation of rare earth metal (III) carbonates under atmospheric pressure.

5. Conclusions

These experimental findings confirm the existence of complex compounds of rare earth metals in the form of ion-associates within a high-concentration carbonate-ion medium (3 mol/L). The strength of electrostatic attraction between REM (III) ions and carbonate anions systematically increased from the light to the heavy REM group, which directly influenced the stability of the resulting associates. Ion-associates of light REMs and medium-mass REMs (up to terbium) were predominantly characterized as loose, whereas those formed by metals beyond terbium were more likely to be contact associates. Variations in the time required for REM (III) carbonates to precipitate resulted in distinct concentration thresholds for different REM groups. These observations suggest that during the carbonate–alkaline processing of primary or secondary REM sources into REM (III) carbonates, the formation of ion-associates as intermediate products may facilitate group separation of REMs through selective precipitation.

Particularly, phosphogypsum processing by the carbonate method should involve treatment of the solid phase by ammonium carbonate exclusively or a mixed ammonium and potassium (sodium) carbonates solution, depending on the REM content in the material. Because the vast majority of technogenic PG includes predominantly light REMs, the second variant would be more prospective. In this case, REM ion-associates needed insignificant carbonate ion decrease for decay into REM carbonates.

Author Contributions: Conceptualization, T.L. and S.G.; methodology, T.L.; software, V.S.; validation, V.S. and E.L.; formal analysis, S.G. and E.L.; investigation, S.G.; resources, V.S.; data curation, T.L.; writing—original draft preparation, S.G.; writing—review and editing, T.L. and S.G.; visualization, V.S.; supervision, T.L.; project administration, S.G. All authors have read and agreed to the published version of the manuscript.

Funding: This research received no external funding.

Data Availability Statement: The original contributions presented in this study are included in the article. Further inquiries can be directed to the corresponding author.

Conflicts of Interest: The authors declare no conflicts of interest.

References

1. Cherepovitsyn, A.; Solovyova, V.; Dmitrieva, D. New challenges for the sustainable development of the rare-earth metals sector in Russia: Transforming industrial policies. *Resour. Policy* **2023**, *81*, 103347. [CrossRef]
2. Cuadros-Muñoz, J.-R.; Jimber-del-Río, J.-A.; Sorhegui-Ortega, R.; Zea-De la Torre, M.; Vergara-Romero, A. Contribution of Rare Earth Elements Is Key to the Economy of the Future. *Land* **2024**, *13*, 1220. [CrossRef]
3. Pathapati, S.V.S.H.; Free, M.L.; Sarswat, P.K. A Comparative Study on Recent Developments for Individual Rare Earth Elements Separation. *Processes* **2023**, *11*, 2070. [CrossRef]
4. Leiva, C.; Arroyo-Torralvo, F.; Luna-Galiano, Y.; Villegas, R.; Vilches, L.F.; Fernández Pereira, C. Valorization of Bayer Red Mud in a Circular Economy Process: Valuable Metals Recovery and Further Brick Manufacture. *Processes* **2022**, *10*, 2367. [CrossRef]
5. Saidakhmetov, P.; Piyanzina, I.; Faskhutdinova, A.; Nedopekin, O.; Adyrbekova, G.; Baiman, G.; Suyerkulova, Z.; Romanova, I. Ab Initio Magnetic Properties Simulation of Nanoparticles Based on Rare Earth Trifluorides REF_3 (RE = Tb, Dy, Ho). *Crystals* **2023**, *13*, 1487. [CrossRef]
6. Castro, L.; Blázquez, M.L.; González, F.; Muñoz, J.Á. Biohydrometallurgy for Rare Earth Elements Recovery from Industrial Wastes. *Molecules* **2021**, *26*, 6200. [CrossRef]
7. Pirzada, M.D.S. Alternative Resources of Rare Earth Elements in Pakistan. *Mater. Proc.* **2024**, *17*, 26. [CrossRef]
8. Amirshahi, S.; Jorjani, E. Preliminary Flowsheet Development for Mixed Rare Earth Elements Production from Apatite Leaching Aqueous Solution Using Biosorption and Precipitation. *Minerals* **2023**, *13*, 909. [CrossRef]
9. Cook, N.J.; Ciobanu, C.L.; Wade, B.P.; Gilbert, S.E.; Alford, R. Mineralogy and Distribution of REE in Oxidised Ores of the Mount Weld Laterite Deposit, Western Australia. *Minerals* **2023**, *13*, 656. [CrossRef]
10. Mitrofanova, G.V.; Chernousenko, E.V.; Kompanchenko, A.A.; Kalugin, A.I. Specific action of collector from phosphoric acid alkyl esters class in flotation of apatite-nepheline ores. *J. Min. Inst.* **2024**, *268*, 637–645.
11. Skublov, S.G.; Levashova, E.V.; Mamykina, M.E.; Gusev, N.I.; Gusev, A.I. The polyphase Belokurikhinsky granite massif, Gorny Altai: Isotope-geochemical study of zircon. *J. Min. Inst.* **2024**, *268*, 552–575.
12. Pashkevich, M.A.; Danilov, A.S. Ecological security and sustainability. *J. Min. Inst.* **2023**, *260*, 153–154.
13. Ma, F.; Chen, L.; Lin, Z.; Liu, Z.; Zhang, W.; Guo, R. Microstructure and Key Properties of Phosphogypsum-Red Mud-Slag Composite Cementitious Materials. *Materials* **2022**, *15*, 6096. [CrossRef] [PubMed]
14. Prischepa, O.M.; Kireev, S.B.; Nefedov, Y.V.; Martynov, A.V.; Lutsky, D.S.; Krykova, T.N.; Sinitsa, N.; Xu, R. Theoretical and methodological approaches to identifying deep accumulations of oil and gas in oil and gas basins of the Russian Federation. *Front. Earth Sci.* **2023**, *11*, 1192051. [CrossRef]
15. Mukaba, J.-L.; Eze, C.P.; Pereao, O.; Petrik, L.F. Rare Earths' Recovery from Phosphogypsum: An Overview on Direct and Indirect Leaching Techniques. *Minerals* **2021**, *11*, 1051. [CrossRef]
16. Chernysh, Y.; Yakhnenko, O.; Chubur, V.; Roubík, H. Phosphogypsum Recycling: A Review of Environmental Issues, Current Trends, and Prospects. *Appl. Sci.* **2021**, *11*, 1575. [CrossRef]
17. Elbagory, M.; Shaker, E.M.; El-Nahrawy, S.; Omara, A.E.-D.; Khalifa, T.H. The Concurrent Application of Phosphogypsum and Modified Biochar as Soil Amendments Influence Sandy Soil Quality and Wheat Productivity. *Plants* **2024**, *13*, 1492. [CrossRef]
18. Guan, Q.; Sui, Y.; Liu, C.; Wang, Y.; Zeng, C.; Yu, W.; Gao, Z.; Zang, Z.; Chi, R.-a. Characterization and Leaching Kinetics of Rare Earth Elements from Phosphogypsum in Hydrochloric Acid. *Minerals* **2022**, *12*, 703. [CrossRef]
19. Ziying, C.; Zhan, L.; Jia, C.; Parashuram, K.; Fawzi, B.; Hongdeng, Q. Recent advances in selective separation technologies of rare earth elements: A review. *J. Environ. Chem. Eng.* **2022**, *10*, 107104.
20. Cheremisina, O.V.; Gorbacheva, A.A.; Balandinsky, D.A.; Luo, Y.; Ponomareva, M.A. Synergistic effect of a mixture of ethoxyphosphoric esters and sodium oleate in aqueous solutions. *Colloids Surf. A Physicochem. Eng. Asp.* **2024**, *685*, 133314. [CrossRef]
21. Pyagai, I.; Zubkova, O.; Babykin, R.; Toropchina, M.; Fediuk, R. Influence of Impurities on the Process of Obtaining Calcium Carbonate during the Processing of Phosphogypsum. *Materials* **2022**, *15*, 4335. [CrossRef] [PubMed]
22. Lutskiy, D.S.; Lukyantseva, E.S.; Mikheeva, V.Y.; Grigorieva, L.V. Investigation of the extraction of samarium and gadolinium from leaching solutions of phosphorus-containing raw materials using solid extractants. *Arab J. Basic Appl. Sci.* **2023**, *30*, 68–73. [CrossRef]
23. Millero, F.J.; Schreiber, D.R. Use of the ion pairing model to estimate activity coefficients of the ionic components of natural waters. *Am. J. Sci.* **1982**, *282*, 1508–1540. [CrossRef]

24. Byrne, R.H. Inorganic speciation of dissolved elements in seawater: The influence of pH on concentration ratios. *Geochem. Trans.* **2003**, *3*, 11–16. [CrossRef] [PubMed]
25. Sobolev, B.P.; Sulyanova, E.A. Lanthanide Contraction in LnF_3 (Ln = Ce-Lu) and Its Chemical and Structural Consequences: Part 2: Specialized Empirical System of R^{3+} (R = Y, La, and 14 Ln) and F^{1-} Ionic Radii for RF_3 Series. *Int. J. Mol. Sci.* **2023**, *24*, 17080. [CrossRef]
26. Sukhanova, K.G. Trace elements in the silicate minerals of the Borodino Meteorite (H5). *J. Min. Inst.* **2024**, *265*, 16–33.
27. Lan, Q.; Yang, Y.; Xie, Z.; Guo, H.; Liu, D.; Zhang, X. Molecular Dynamics Calculation of the Coordination Behavior of Yb (III) in Sodium Carbonate Solution. *Processes* **2023**, *11*, 2624. [CrossRef]
28. Surkova, A.; Bogomolov, A.; Paderina, A.; Khistiaeva, V.; Boichenko, E.; Grachova, E.; Kirsanov, D. Milk Analysis using a New Optical Multisensor System Based on Lanthanide (III) Complexes. *Eng. Proc.* **2023**, *48*, 28. [CrossRef]
29. Cantrell, K.; Byrne, R. Rare earth element complexation by carbonate and oxalate ions. *Geochim. Cosmochim. Acta* **1987**, *51*, 597–605. [CrossRef]
30. Lobacheva, O.L. Ion Flotation of Ytterbium Water-Salt Systems—An Innovative Aspect of the Modern Industry. *Water* **2021**, *13*, 3493. [CrossRef]
31. Firsching, F.; Mohammadzadei, J. Solubility products of the rare-earth carbonates. *J. Chem. Eng. Data* **1986**, *31*, 40–42. [CrossRef]
32. Ohta, A. Experimental and theoretical studies of REE partitioning between Fe hydroxide and Mn oxide and seawater. *Chikyukagaku (Geochem.)* **2006**, *40*, 13–30, (In Japanese with English abstract).
33. Olender, R.; Nitzan, A. Lattice theory of solvation and dissociation in macromolecular fluids. II. Quasichemical approximation. *J. Chem. Phys.* **1994**, *101*, 2338–2349. [CrossRef]
34. Finney, A.R.; Lectez, S.; Freeman, C.L.; Harding, J.H.; Stackhouse, S. Ion Association in Lanthanide Chloride Solutions. *Chem. A Eur. J.* **2019**, *25*, 8725–8740. [CrossRef] [PubMed]
35. Smolyakov, B.; Veselova, G. Limiting Equivalent Conductivities of Ions in Water Between 25 and 200 C. *Elektrokhim* **1975**, *11*, 700–704.
36. Ismail, I.M.; Masuko, Y.; Tomiyasu, H.; Fujii, Y. Conductivity of some rare earth chlorides in sub and supercritical aqueous solutions. *J. Supercrit. Fluids* **2003**, *25*, 69–79. [CrossRef]
37. Kellermeier, M.; Raiteri, P.; Berg, J.K.; Kempter, A.; Gale, J.D.; Gebauer, D. Entropy drives calcium carbonate ion association. *ChemPhysChem* **2016**, *17*, 3535–3541. [CrossRef]
38. Oleynik, I.L. Increasing the Processing Degree of Phosphate Raw Materials with the Associated Extraction of the Rare Earth Metals. The Dissertation Abstract for the Degree of Candidate of Technical Sciences. 2021. Available online: https://spmi.ru/sites/default/files/imci_images/sciens/dissertacii/2021/oleynik_avtoreferat.pdf (accessed on 17 February 2025).
39. Pawlowicz, R. Calculating the conductivity of natural waters. *Limnol. Oceanogr. Methods* **2008**, *6*, 489–501. [CrossRef]
40. Fang, Y. Ions Interacting with Macromolecules NMR Studies in Solution. Ph.D. Thesis, KTH Royal Institute of Technology, Stockholm, Sweden, 2017; p. 71.
41. Litvinova, T.E.; Oleinic, I.S.; Lutskiy, D.S. Thermodynamic justification of the extraction of rare-earth metals during the carbonate conversion of secondary phosphate raw materials. *ARPN J. Eng. Appl. Sci.* **2020**, *15*, 2919–2924.

metals

Article

Microwave Treatment of Copper–Nickel Sulfide Ore for Promotion of Grinding and Flotation

Xiaolei Fang, Zhiwei Peng *, Tianle Yin, Mingjun Rao and Guanghui Li

School of Minerals Processing and Bioengineering, Central South University, Changsha 410083, China
* Correspondence: zwpeng@csu.edu.cn; Tel.: +86-731-88877656

Abstract: The effect of microwave treatment on the grinding and flotation performance of a typical copper–nickel sulfide ore was evaluated, based on the determination of its microwave absorption capability, grinding and flotation indexes such as crack percentage, mineral liberation degree, particle size distribution, relative work index (RWI), metal enrichment ratio and recovery. There were obvious differences between the microwave absorption capabilities of the main minerals in the ore, as demonstrated by their different microwave penetration depths. They also induced temperature differences between sulfide minerals and gangue minerals which could reach 418 °C after microwave treatment for 20 s. It was shown that microwave treatment could effectively improve the grindability of the ore, as proven by the increase in fine particles smaller than 0.074 mm and the decrease in RWI after grinding due to the higher crack percentage and mineral liberation degree. Moreover, microwave treatment affected the ore floatability because of the generation of cuprite, retgersite, and rozenite with poor floatability when the treatment time was extended. By microwave treatment for a proper time, 20 s, an optimal balance between the grindability and flotation performance could be achieved. Compared with the untreated ore, the RWI of the ore decreased by 11.5%. After flotation, the Cu and Ni enrichment ratios of the flotation concentrate increased by 0.3 and 0.2, respectively. Meanwhile, their corresponding recoveries increased by 4.2% and 3.1%. This study provides new insights for the treatment of copper–nickel sulfide ore to enhance the grinding and flotation process.

Keywords: microwave treatment; copper–nickel sulfide ore; thermodynamics; grinding indexes; flotation

1. Introduction

Nickel is an important strategic metal with high temperature and corrosion resistances, high ferromagnetism, and good plasticity [1]. It has important applications in stainless steel, batteries, high temperature alloys, catalysts, etc. [2,3]. The current sources of nickel ore for industrial applications in the world are nickel sulfide ore and nickel laterite ore [4]. The Jinchuan deposit is the third largest magmatic copper–nickel sulfide deposit in the world and the largest one in China [5]. In general, increasing the grade of nickel sulfide ore for smelting is important for the full utilization of the resource [6]. With the continuous mining of nickel sulfide ore, the proportion of low-grade ore keeps increasing [7]. To meet the smelting requirements of nickel sulfide ore, a preliminary mineral concentration by flotation is often required [8]. Before the flotation, it is necessary to improve the mineral liberation degree of the ore by crushing and/or grinding to improve the flotation efficiency [9,10]. It was reported that the energy consumption during crushing and grinding occupies 50–70% of the whole process, while the energy utilization efficiency is less than 1% and more than 90% of the energy is lost in the form of heat, noise, and so on [11–13]. Therefore, it is critical to develop more efficient and sustainable grinding processes.

Microwave is a type of electromagnetic wave having the frequency within the range 300 MHz–300 GHz. It has the selective and volumetric thermal effects which depend on the microwave absorption capabilities of target materials. It is widely applied for heating to save energy consumption with environmental friendliness, which remains a big challenge for conventional methods. Currently, microwave radiation had been used in grinding [9,10], drying [14], metallurgy [15], materials synthesis and processing [16,17], and other fields [18,19]. For metal ores, the different microwave absorption capabilities between valuable minerals and gangue minerals lead to dissimilar heating responses, which will create internal stresses and subsequently cracks in the ore in a very short time, i.e., in seconds in many cases [20,21]. These cracks will strengthen the grinding performance of the ore with lower energy consumption by improving mineral liberation. For a copper ore with the grade of 1.7–1.8% after exposure to microwave radiation for 30 s, its grinding time could be reduced by 36.9% and the particle size at 80% of the cumulative distribution (D_{80}) after grinding decreased from 2.00 mm to 1.29 mm [21]. For chalcopyrite ore, its microwave treatment and microflotation test showed that the recovery of chalcopyrite increased from 41% to 51% after microwave treatment for 5 s and decreased to 16% and 4% after the treatment for 10 s and 20 s, respectively, because prolonged microwave treatment time might cause the formation of hydrophilic iron oxides/hydroxides/oxyhydroxides and sulfate species at the chalcopyrite surface, which significantly reduced floatability of the ore [22,23].

Considering the higher processing cost of laterite ore than nickel sulfide ore [24], it is valuable to improve the efficiency of grinding and flotation for strengthening the utilization of nickel sulfide ore. By now, few studies have been conducted on the Jinchuan copper–nickel sulfide ore by focusing on the effect of microwave treatment on the cracking, mineral liberation, and phase composition of the ore [25]. In this study, the microwave absorption capabilities and heating rates of the main minerals in a typical complex copper–nickel sulfide ore in the Jinchuan deposit were investigated. In addition, the effects and mechanisms of microwave treatment on the grinding and flotation characteristics of the ore were evaluated.

2. Experimental

2.1. Raw Materials

The copper–nickel sulfide ore used in this study was taken from the Jinchuan area, Jinchang, China. As shown in Table 1, the contents of Cu and Ni, determined by inductively coupled plasma optical emission spectroscopy (ICP-OES), were 1.24 wt% and 1.27 wt%, respectively. The total contents of other components, mainly MgO, SiO_2, Fe, and S, which were measured by X-ray fluorescence (XRF), exceeded 95 wt%.

Table 1. Main chemical composition of copper–nickel sulfide ore (wt%).

Component	MgO	SiO_2	Al_2O_3	CaO	Fe	S	Cu	Ni
Content	30.26	32.60	2.68	2.50	24.22	9.25	1.24	1.27

Figure 1 shows the XRD pattern of the ore. The main phases of copper–nickel sulfide ore used in this study were lizardite ($Mg_3[(Si,Fe)_2O_5](OH)_4$), cordierite ($Mg_2Al_4Si_5O_{18}$), quartz (SiO_2), amesite ($Mg_2Al[(Si,Al)_2O_5](OH)_4$), chalcopyrite ($CuFeS_2$), pentlandite ($(Fe,Ni)_9S_8$), pyrrhotite ($Fe_7S_8$), and pyrite ($FeS_2$). Its elemental distribution was determined by scanning electron microscopy (SEM)—energy-dispersive spectrometry (EDS) analysis. As shown in Figure 2, Cu, Ni, Fe, and S were the main elements of the target sulfide minerals, i.e., chalcopyrite and pentlandite, while Mg and Si were the primary elements of the gangue minerals. The valuable minerals were tightly bound and wrapped by gangue minerals, indicating the necessity of sufficient ore liberation for subsequent processing. For comparison and microwave heating mechanism analysis, various metal sulfide mineral samples, including chalcopyrite, pyrite, and pyrrhotite, were collected. Figure 3 shows their XRD patterns, which proved their high purity.

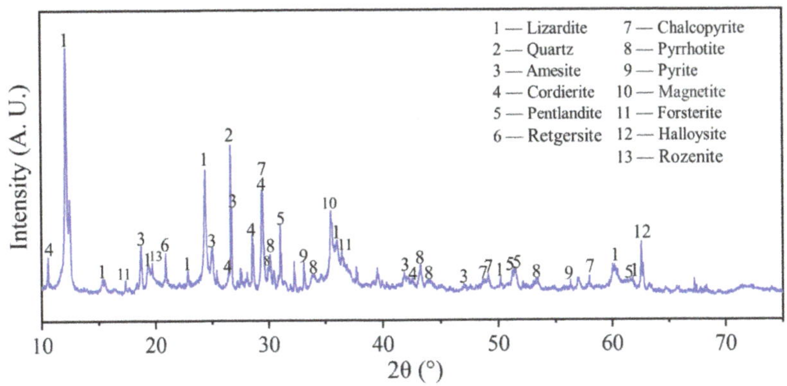

Figure 1. XRD pattern of copper–nickel sulfide ore.

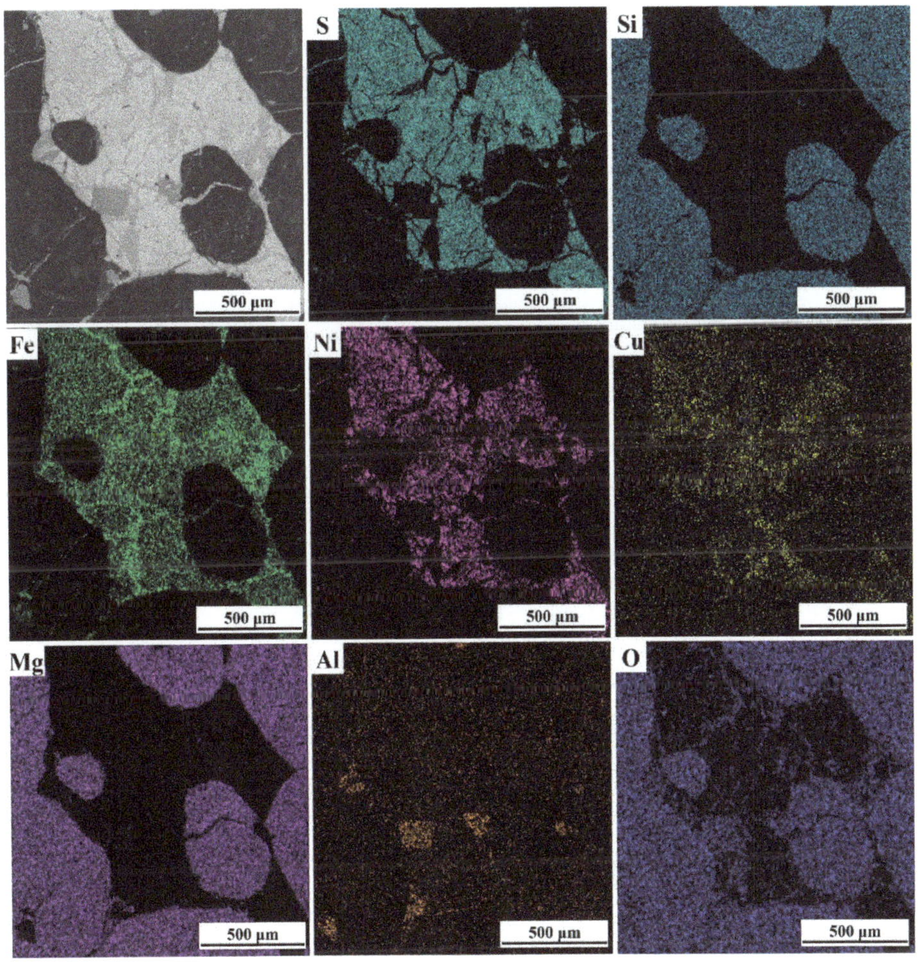

Figure 2. SEM-EDS analysis of copper–nickel sulfide ore.

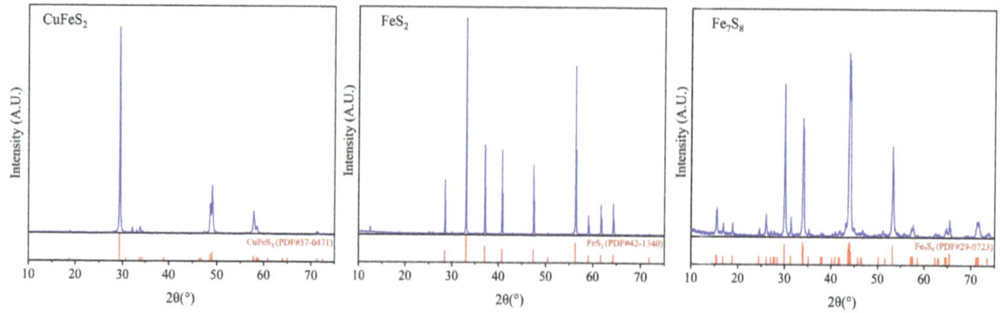

Figure 3. XRD patterns of chalcopyrite, pyrite, and pyrrhotite.

2.2. Experimental Procedure

2.2.1. Thermodynamic Calculation

The variations of standard Gibbs free energy changes of possible reactions ($\Delta_r G_m^{\theta}$) with temperature (T) during the microwave treatment process of the copper–nickel sulfide ore were calculated using the software FactSage 8.1 (Thermfact/CRCT, Montreal, QC, Canada; GTT-Technologies, Herzogenrath, Germany) to evaluate the feasibility of the reactions.

2.2.2. Microwave Absorption Capability Determination

The permittivities and permeabilities of the ore and the main sulfide minerals, including chalcopyrite, pyrite, and pyrrhotite, were measured firstly by the coaxial method using a vector network analyzer (VNA, E5071C, Agilent Technology Co., Ltd., Santa Clara, CA, USA) in the frequency range 2–4 GHz. Note that pentlandite could be intimated by pyrrhotite with similar properties and the magnetism of chalcopyrite, pyrite, and pyrrhotite would also be beneficial for microwave treatment of the ore, as reported in the literature [20,26–29]. Also, pure gangue minerals related to those in the ore were not prepared for the same measurement because their properties have been reported in detail [8,30–32]. For the measurements, the sample particles with sizes of −0.074 mm were mixed with paraffin wax in a mass ratio of 7/3 and pressed through a pressing machine to make ring samples with an inner diameter of 3 mm, an outer diameter of 7 mm, and a thickness of 2 mm. By using the measured values, the microwave absorption capabilities of the materials could be determined by calculating the corresponding values of microwave penetration depth, D_p, which is defined as the depth at which the microwave power falls to 1/e (e: natural constant) of its original or surface value. Its equation is given as follows [33]:

$$D_p = \frac{\lambda_0}{2\sqrt{2}\pi}\left\{\varepsilon_r''\mu_r'' - \varepsilon_r'\mu_r' + \left[(\varepsilon_r'\mu_r')^2 + (\varepsilon_r''\mu_r'')^2 + (\varepsilon_r'\mu_r'')^2 + (\varepsilon_r''\mu_r')^2\right]^{\frac{1}{2}}\right\}^{-\frac{1}{2}} \quad (1)$$

where λ_0 is the microwave wavelength in free space; ε_r' is the real part of the complex relative permittivity of the material (relative dielectric constant); ε_r'' is the imaginary part of complex relative permittivity of the material (relative dielectric loss factor); μ_r' is the real part of complex relative permeability of the material (relative magnetic constant); and μ_r'' is the imaginary part of the complex relative permeability of the material (relative magnetic loss factor).

2.2.3. Microwave Treatment

Microwave treatment was carried out in a microwave tube furnace (CY-SA1700C-S, Hunan Changyi Microwave Technology Co., Ltd., Changsha, China). By considering the working conditions of the furnace and preliminary experimental results, in each test, 120 g of the ore with a particle size of −1.25 + 0.60 mm was used and placed in a microwave-transparent corundum crucible in the furnace for treatment. The microwave frequency and power were 2.45 GHz and 2200 W, respectively, and the treatment time was set at 0 s,

10 s, 20 s, 30 s, and 40 s, respectively. The temperatures of the samples during microwave treatment were measured by a thermocouple (53IIB, FLUKE Testing Instruments Co., Ltd., Everett, WA, USA), which was inserted inside the samples. After the treatment, the samples were taken out and naturally cooled for characterization and subsequent experiments. The phase compositions of the samples were determined by an X-ray diffraction spectrometer (XRD, D/max 2550PC, Rigaku Co., Ltd., Tokyo, Japan) using a Cu-anode target with the wavelength of 1.54056 Å, step scan mode, and step length of 0.02°. For micro-morphology analysis, the samples were embedded in epoxy glue for fixation and polishing. The microstructures and elemental distributions of the samples were determined by a scanning electron microscope (SEM, MIRA3 LMH, TESCAN, Brno, Czech Republic) equipped with an energy-dispersive X-ray spectrometer (EDS, One Max 20, Oxford Instruments PLC, Oxford, UK). The cracks in the microstructure images were identified and binarized for calculating the percentage of light area as the crack percentage using the Advanced Weka Segmentation module in the software ImageJ 2.0 (National Institutes of Health, Bethesda, MD, USA).

2.2.4. Grinding and Flotation

For grinding, 110 g of microwave-treated ore was ground in a ball mill with a liquid-to-solid ratio of 1 for 14 min. The particle size distributions of the pulps after grinding were determined using a laser particle size analyzer (Malvern Mastersizer 2000, Malvern Panalytical, Malvern, UK). The ground samples were embedded with epoxy resin and polished, as observed and photographed from one field of view after another using an optical microscope (DMI5000M, Leica Instruments Co., Ltd., Mannheim, Germany). Afterwards, according to the area occupied by the target minerals in the ore particles under the microscope which could be classified into five types: 1/5, 2/5, 3/5, 4/5, 1, denoted by $N_{1/5}$, $N_{2/5}$, $N_{3/5}$, $N_{4/5}$, N_1, respectively, the mineral liberation degree (L) was calculated using the following equation.

$$L = \frac{N_1}{N_1 + \frac{1}{5}N_{1/5} + \frac{2}{5}N_{2/5} + \frac{3}{5}N_{3/5} + \frac{4}{5}N_{4/5}} \times 100\% \tag{2}$$

After grinding, the pulp was left to settle for 50 min. After removing the top layer of the clear liquid, the bottom pulp was moved to the flotation cell while the flotation liquid-to-solid ratio was adjusted. Pulp sampling was carried out in the flotation cell using a sampler and multiple pulp sampling was carried out at different locations in the cell to ensure the representativeness of the removed pulp. Before the laser particle size analysis, sodium hexametaphosphate ($(NaPO_3)_6$) was added as a dispersant in the pulp and subjected to ultrasonic treatment for 5 min to avoid the serious agglomeration of the gangue mineral, namely lizardite [5]. To assess the grindability of the ore, the values of relative work index (RWI) were calculated using the following equation [8]:

$$RWI = \left[\left(\frac{10}{\sqrt{P_r}} - \frac{10}{\sqrt{F_r}}\right) \Big/ \left(\frac{10}{\sqrt{P_t}} - \frac{10}{\sqrt{F_t}}\right)\right] \times 100\% \tag{3}$$

where P and F denote the 80% passing sizes of the grinding product and feed ore, respectively; and r and t denote the untreated and treated ores, respectively. A RWI less than 100% indicates an increase in ore grindability and a decrease in ore particle size, while a RWI greater than 100% shows a decrease in ore grindability and an increase in ore size.

The flotation experiments were carried out in a 0.5 L flotation cell with a rotation speed of 2000 rpm and a pulp concentration of 20% during the flotation conditioning stage (XFD IV, Jilin Exploration Machinery Company, Jilin, China). The automatic froth scraping was carried out with an air flow rate of 3.2 L/min and a scraper speed of 30 rpm. Sodium hexametaphosphate ($(NaPO_3)_6$), copper sulfate ($CuSO_4$), butyl sodium xanthate ($C_4H_9OCSSNa$), and terpineol were used as the depressant, activator, collector, and frother, respectively. The pulp pH was 7. Figure 4 shows the flotation flowsheet. It included one rougher flotation stage for 30 s and four scavenger flotation stages for 30 s, 60 s, 60 s and 60 s, respectively. The corresponding concentrate and middlings were named K, Π_1, Π_2, Π_3, and Π_4, respectively. Correspondingly, the flotation tailing was named X.

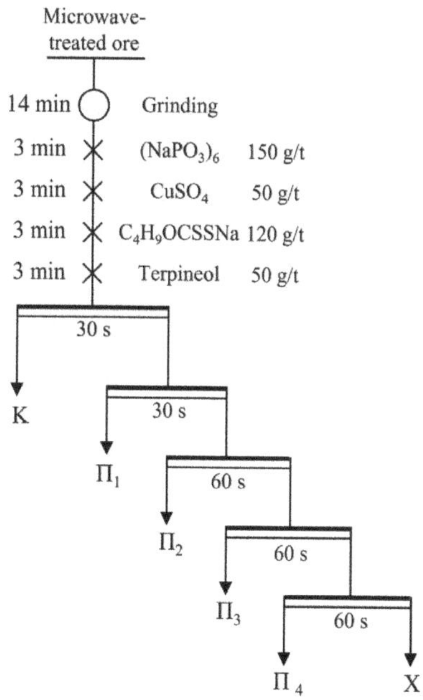

Figure 4. Flotation flowsheet.

The Cu and Ni contents of the samples were measured using an inductively coupled plasma–optical emission spectrometer (ICP-OES, Optima 5300, PerkinElmer, Waltham, MA, USA) before which 0.1000 ± 0.0010 g of each sample was dissolved with aqua regia (volume ratio, HCl/HNO_3 = 3, 30 mL), HF (4 mL), and $HClO_4$ (4 mL) by heating at 140 °C for 30 min and then at 220 °C for 30 min. The metal enrichment ratios, *E*, and recoveries, *R*, of the flotation concentrate, were calculated by Equations (4) and (5), respectively:

$$E = \frac{c}{f} \tag{4}$$

$$R = \frac{\gamma \times c}{f} \times 100\% \tag{5}$$

where *c* is the copper or nickel content in the concentrate; *f* is the copper or nickel content of copper–nickel sulfide ore, and γ is the yield of the concentrate.

The flotation data were fitted using a first-order kinetic model with rectangular distribution of floatability to further investigate the effect of microwave treatment on the flotation behavior [34]. The related equation is given as follows:

$$R = R_\infty \left[1 - \frac{1}{kt}\left(1 - e^{-kt}\right)\right] \tag{6}$$

where *R* is the recovery at the cumulative flotation time *t*; R_∞ is the theoretical maximum recovery; and *k* is the first-order flotation rate constant.

3. Results and Discussion

3.1. Thermodynamic Analysis

To determine the effect of microwave treatment on the phase transformation of the valuable minerals in the ore, the variations of the standard Gibbs free energy changes of the relevant reactions of chalcopyrite, pyrite and pyrrhotite with temperature were determined in the temperature range from 25 °C to 800 °C. Due to the complexity of oxidation behaviors of pentlandite, $(Ni,Fe)_9S_8$, it cannot be calculated directly. Instead, its main preliminary oxidation products, NiS and FeS, were considered [35]. The results are shown in Table 2 and Figure 5. Considering the negative values of $\Delta_rG_m^\theta$ of the reactions at low temperatures, the oxidations of Fe_7S_8, $CuFeS_2$, FeS_2, NiS, and their weak oxidation products (FeS, Cu_2S, etc.) are thermodynamically feasible. The oxidation of $FeSO_4$ and the decomposition of $NiSO_4$ may also occur by elevating temperature.

Table 2. Standard Gibbs free energy changes of the main reactions involved in the treatment of copper–nickel sulfide ore.

No.	Reaction Equation	$\Delta_rG_m^\theta$ − T (kJ/mol)
(7)	$2CuFeS_2(s) + 4O_2(g) = 2FeO(s) + Cu_2S(s) + 3SO_2(g)$	−1117.35 + 0.098T
(8)	$2Cu_2S(s) + 3O_2(g) = 2Cu_2O(s) + 2SO_2(g)$	−708.26 + 0.204T
(9)	$2Cu_2O(s) + O_2(g) = 4CuO(s)$	−220.49 + 0.202T
(10)	$4CuFeS_2(s) + 13O_2(g) = 4CuO(s) + 2Fe_2O_3(s) + 8SO_2(g)$	−3676.19 + 0.859T
(11)	$4Fe_7S_8(s) + 9O_2(g) = 6Fe_2O_3(s) + 16FeS_2(s)$	−4033.94 + 2.619T
(12)	$FeS_2(s) + O_2(g) = FeS(s) + SO_2(g)$	−242.56 −0.064T
(13)	$FeS(s) + 2O_2(g) = FeSO_4(s)$	−729.23 + 0.358T
(14)	$4FeSO_4(s) + O_2(g) = 2Fe_2O_3(s) + 4SO_3(g)$	335.18 − 0.488T
(15)	$4FeS(s) + 7O_2(g) = 4SO_2 + 2Fe_2O_3(s)$	−2286.74 + 0.564T
(16)	$NiS(s) + 2O_2(g) = NiSO_4(s)$	−675.18 + 0.347T
(17)	$2NiS(s) + 3O_2(g) = 2NiO(s) + 2SO_2(g)$	−846.11 + 0.151T
(18)	$NiSO_4(s) = NiO(s) + SO_3(g)$	178.38 − 0.177T

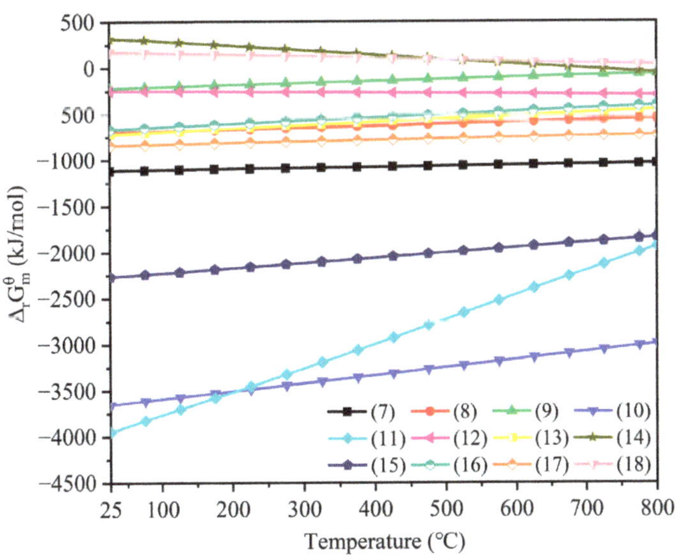

Figure 5. $\Delta_rG_m^\theta$ − T diagram of the main reactions involved in the treatment of copper–nickel sulfide ore.

3.2. Microwave Absorption Capability Analysis

Figure 6 shows the values of the complex relative permittivity and complex relative permeability of copper–nickel sulfide ore, chalcopyrite, pyrite, and pyrrhotite in the frequency range 2.0–4.0 GHz. These parameters remained relatively stable as the frequency varied. Figure 7 shows their D_p values at 2.45 GHz. It was found that the D_p value of copper–nickel sulfide ore was 20.57 mm, higher than those of chalcopyrite, pyrite, and pyrrhotite (all around 15 mm). It indicated the better microwave absorption capabilities of the metal sulfide minerals than that of the ore itself. This was because the ore contained both metal sulfide minerals and gangue minerals, such as magnesium or aluminum silicates, which were considered to be nearly microwave-transparent at room temperature, as verified by the measurements of their electromagnetic properties and heating rates [30–32]. Obviously, due to the significant differences of microwave absorption capabilities between the minerals, they would have different microwave heating characteristics.

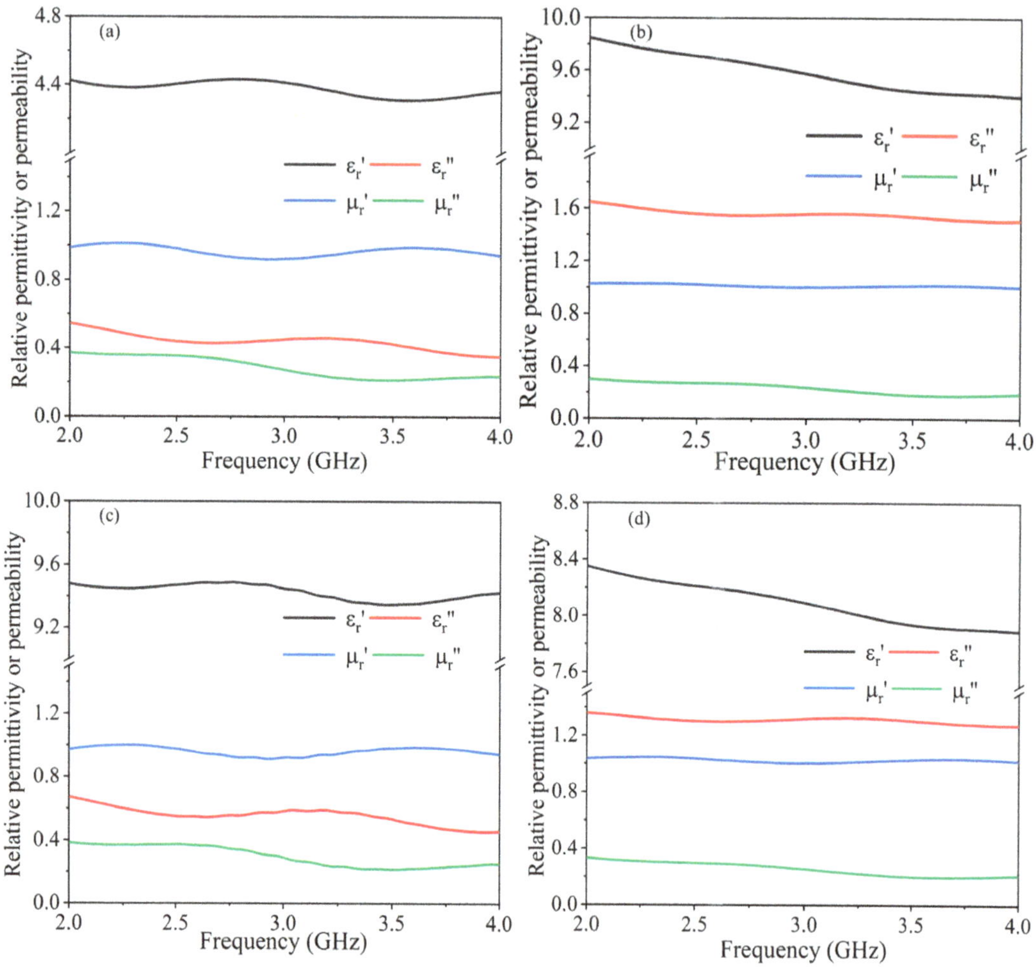

Figure 6. Complex relative permittivities and permeabilities of (**a**) copper–nickel sulfide ore, (**b**) chalcopyrite, (**c**) pyrite, and (**d**) pyrrhotite.

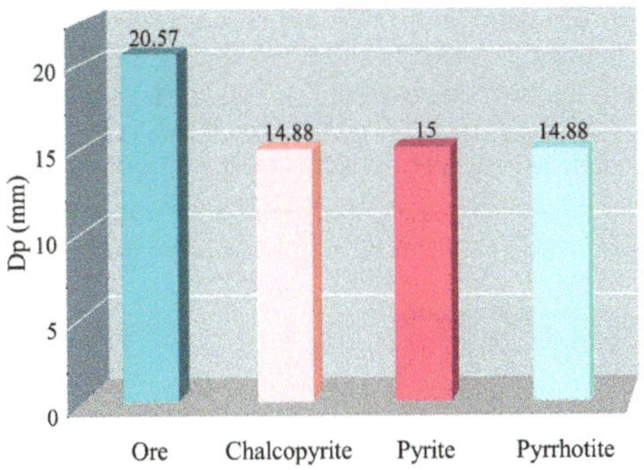

Figure 7. D_p values of copper–nickel sulfide ore and different minerals at 2.45 GHz.

3.3. Microwave Treatment Analysis

3.3.1. Microwave Heating Characteristics

Figure 8 shows the temperature changes of copper–nickel sulfide ore and different minerals as a function of microwave treatment time. As shown in Figure 8, in the early stage of microwave treatment (0–15 s), the differences in temperature between the ore and minerals kept increasing. Pyrite, chalcopyrite, and pyrrhotite had higher heating rates than the ore due to their better microwave absorption capabilities than the ore at low temperature, as proven in Figure 7. After microwave treatment for 15–20 s, their temperature increasing trends became slower due to possible oxidation reactions. It indicated that further extending the treatment time may not be conducive to improving the overall efficiency of grinding and flotation. Unlike the above minerals, the gangue mineral, lizardite, had a lower microwave heating rate. The maximum temperature difference between sulfide minerals and gangue minerals could reach 418 °C (chalcopyrite vs. lizardite) after microwave treatment for 20 s. The significant temperature differences between the minerals under microwave irradiation provided the feasibility of microwave-assisted grinding. For other gangue minerals, including cordierite and amesite, they showed poor heating performance with minor temperature changes, even after 60 s. Obviously, they were nearly microwave-transparent.

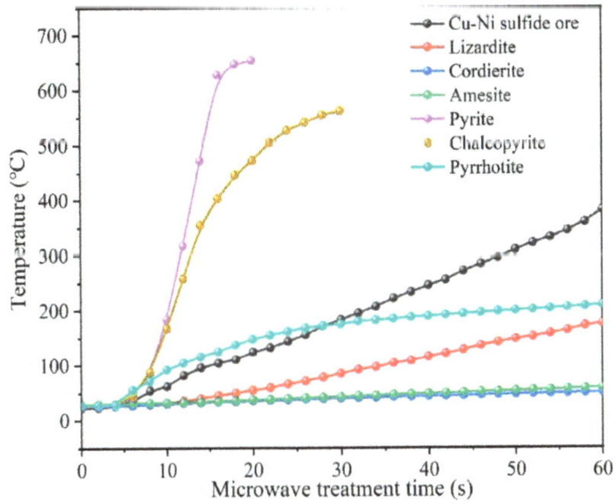

Figure 8. Temperature rising curves of the ore and different minerals during microwave treatment.

Along with the temperature increase during microwave treatment, the structure of the ore would also change. As shown in Figure 9a, obvious macroscopic cracks were produced on the ore surfaces after microwave treatment. Figure 9b shows the microstructure and spot compositions of the ore sample obtained after microwave treatment for 20 s. The obvious cracks appeared on the boundaries of the bright and dark areas and inside them. The main elements in the dark gray area of Figure 9b (spot C) were Mg and Si, indicating the existence of silicates. The light gray area (spot B) was mainly composed of Fe, S, and O, indicating the presence of pyrite and iron oxide. The bright area (spot A) was constituted by Cu, Fe, and S, which were probably the components of chalcopyrite. Figure 10 shows some representative SEM images of internal cracks in the ore after microwave treatment. As the microwave treatment time was extended, the quantity and size of cracks in the ore increased.

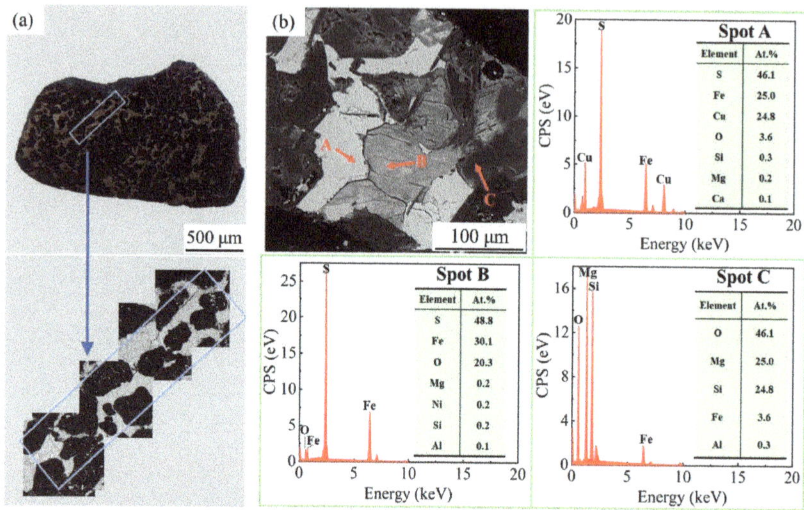

Figure 9. (**a**) Macroscopic cracks in the copper–nickel sulfide ore and (**b**) the microstructure of copper–nickel sulfide ore with EDS analysis.

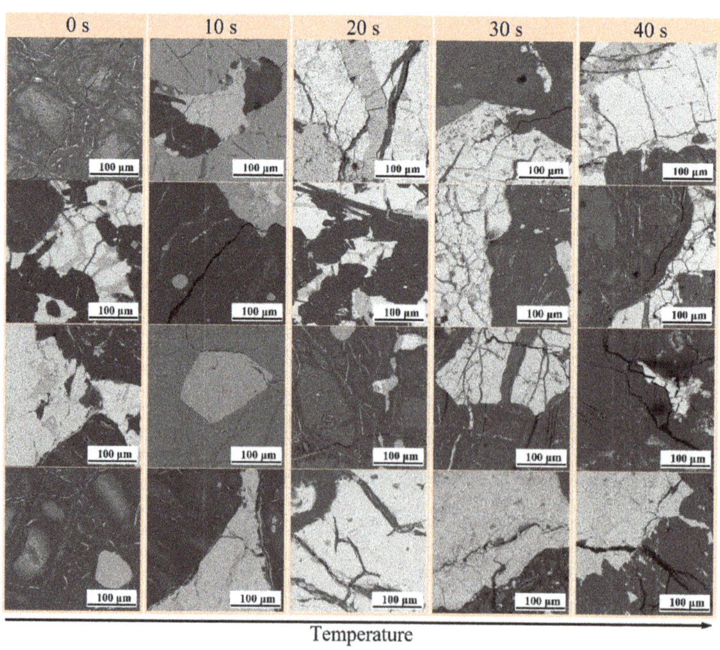

Figure 10. Representative SEM images of the ore after microwave treatment for different time.

3.3.2. Grinding Characteristics of Ore

The crack percentages of the ore samples treated for different time were statistically analyzed. As shown in Figure 11, the crack percentage increased with microwave treatment time. The results indicated that microwave treatment promoted the formation of cracks between different minerals in the ore, as observed from the binarized images of ore cracks after microwave treatment in Figure 12. The cracks would effectively intensify mineral liberation. Figure 13 shows the variation of liberation degree of Cu/Ni-bearing minerals, including chalcopyrite and pentlandite, in the ore with microwave treatment time. It was obvious that the liberation degree increased with the extension of microwave treatment time. However, after microwave treatment for over 20 s, their liberation degrees increased minorly because the quantity of the cracks gradually reached the maximum.

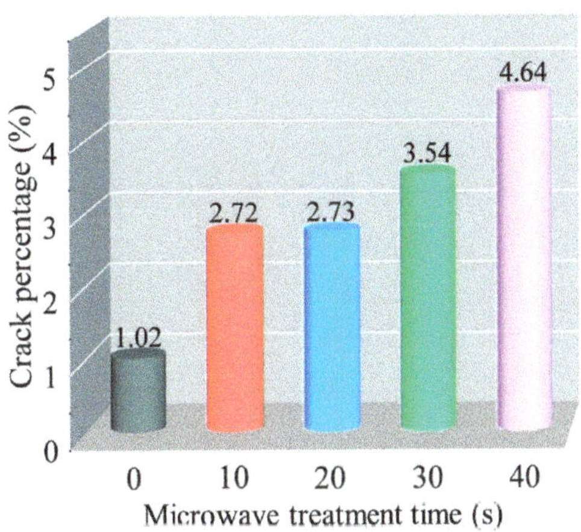

Figure 11. Effect of microwave treatment time on the crack percentage of the ore.

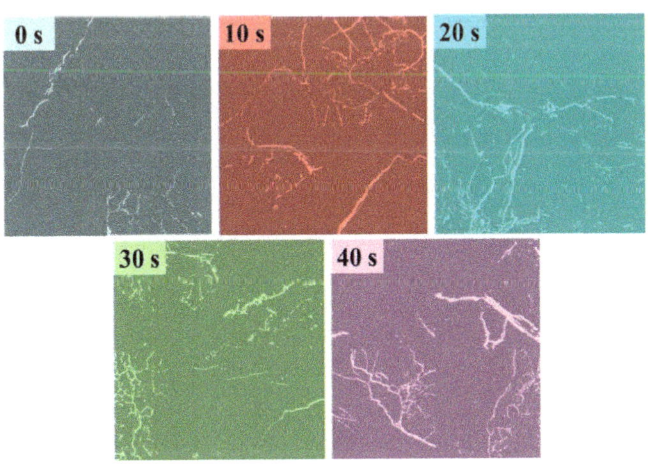

Figure 12. Representative binarized images of ore cracks after microwave treatment.

Figure 14 shows the variation of the particle size distribution of the microwave-treated ore after grinding with microwave treatment time. The particle size differential distribution curves showed two peaks in the size ranges 20–100 μm and 0.1–1 μm. Meanwhile, the cumulative size distribution curve moved to the left with increasing microwave treatment time. Correspondingly, the particle size corresponding to the highest peak of the particle size differential distribution curve decreased while the percentage of −0.074 mm particles increased. These changes indicated that microwave treatment reduced the overall particle size and promoted the particle size uniformity, which was beneficial for mineral liberation. From the curves, the values of D_{10}, D_{50}, and D_{90} (particle sizes at 10%, 50% and 90% of the cumulative distribution, respectively) could be determined, as summarized in Table 3. It was obvious that they kept decreasing with increasing microwave treatment time.

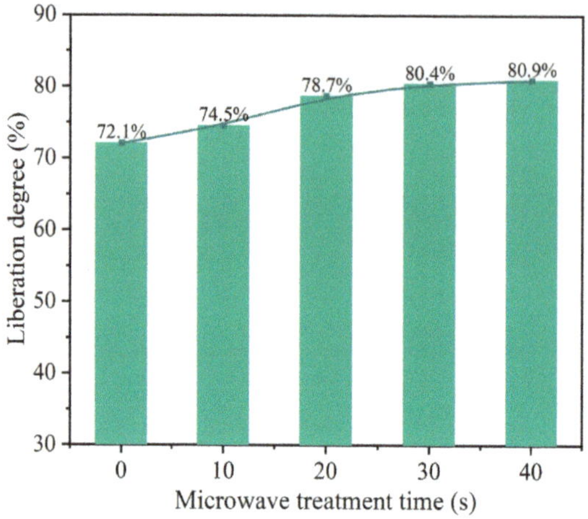

Figure 13. Effect of microwave treatment time on the liberation degree of Cu/Ni-bearing minerals.

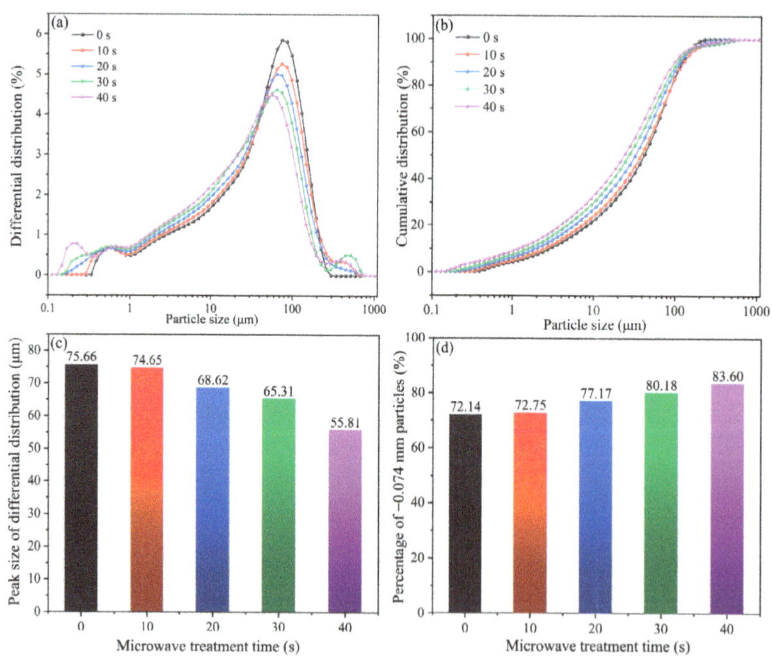

Figure 14. Particle size distribution analysis of the microwave-treated ore after grinding: (**a**) differential distribution, (**b**) cumulative distribution, (**c**) peak sizes determined from differential distribution, and (**d**) percentages of −0.074 mm particles determined from cumulative distribution.

Table 3. Particle size percentages of microwave-treated ore after grinding.

Treatment Time (s)	Particle Size Percentage (%)		
	D_{10}	D_{50}	D_{90}
0	3.011	44.687	126.984
10	2.644	41.252	134.195
20	2.061	34.985	120.777
30	1.673	29.292	112.642
40	1.286	25.469	101.980

It should be noted that the production amount of fine particles in the pulp during grinding should be appropriate. With excessive fine particles, a serious sliming phenomenon would occur, causing the excessive adhesion of gangue minerals to the surface of target minerals. As a result, it would lower the metal recovery. For instance, the heteroaggregation between lizardite and pentlandite would take place [36], reducing the Ni recovery in the concentrate.

The D_{80} and RWI of the ores after microwave treatment for 10 s were almost unchanged compared to the untreated ore (Figure 15). When the treatment time reached 20 s, there were significant decreases in D_{80} and RWI, showing better grindability. In particular, the corresponding RWI decreased by 11.5%. When the treatment time reached 40 s, the RWI decreased to nearly 70% of the untreated sample, which indicated a significant decrease in the grinding energy consumption.

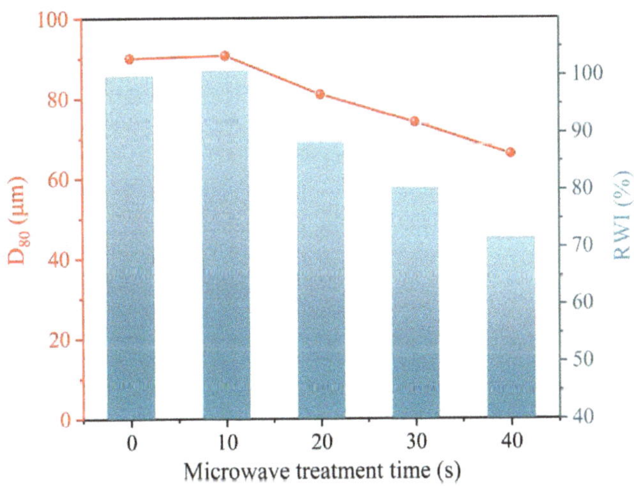

Figure 15. Effect of microwave treatment time on D_{80} and RWI.

3.3.3. Flotation Characteristics of Ground Ore

Figure 16 shows the effect of microwave treatment time on the enrichment ratios and recoveries of Cu and Ni in the flotation concentrate. After microwave treatment for 10 s, both the enrichment ratios and recoveries of Cu and Ni in the concentrate decreased compared to the untreated sample. However, by extending the time to 20 s, the flotation performance improved significantly. The enrichment ratios of Cu and Ni increased by 0.3 and 0.2, respectively, and the recoveries of Cu and Ni increased by 4.2% and 3.1%, respectively, as compared to those of the untreated sample. With a further increase in microwave treatment time, the changes in enrichment ratios and recoveries of Cu and Ni became less considerable. When the treatment time reached 40 s, the recoveries of Cu and Ni decreased by 6.67% and 11.11%, respectively, compared to those of the untreated ore.

Note that the variations of the indexes for Ni were greater than those for Cu due to the stronger heteroaggregation between lizardite and pentlandite [36].

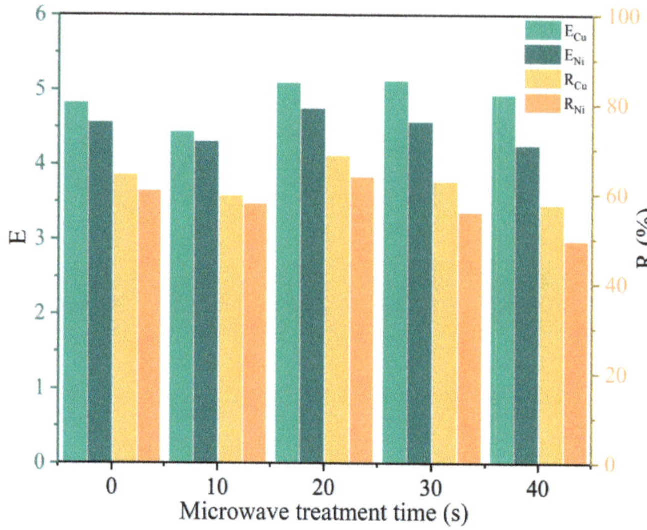

Figure 16. Effect of microwave treatment time on the enrichment ratios and recoveries of Cu and Ni in the flotation concentrate.

Figure 17 shows the flotation data fitting curves from which the corresponding values of k, R_∞, and R^2 (coefficient of determination) were determined. According to Table 4, all R^2 were greater than 0.999 and the results were considered reliable. The changes in R_∞ of Cu and Ni had similar trends after microwave treatment. The value of R_∞ of the ore after microwave treatment for 20 s was higher than that of the untreated ore. When the treatment time exceeded 20 s, the values of both R_∞ and k decreased. It indicated that the oxidation caused by the prolonged microwave treatment degraded the flotation performance. It should be mentioned that the value of k of the microwave-treated ore was smaller than that of the untreated ore in the initial stage of flotation, due to the increase in the content of micro-fine particles in the pulp, in which more gangue mineral particles adhered to the concentrate surface.

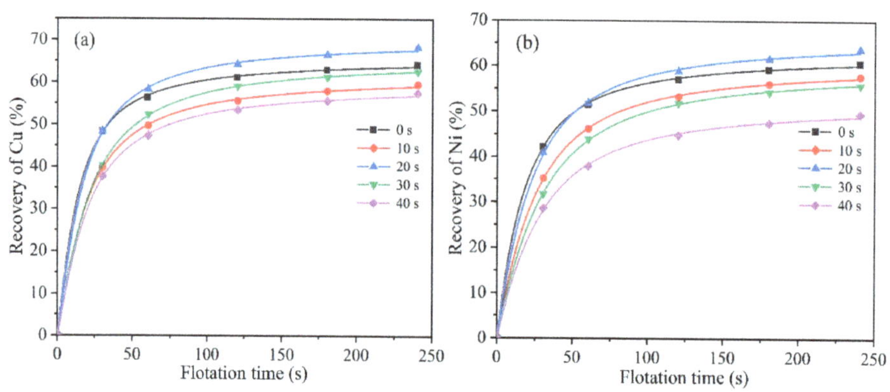

Figure 17. Fitting of flotation data using the first-order kinetic model with rectangular distribution of floatability: (**a**) recovery of Cu and (**b**) recovery of Ni.

Table 4. Fitting parameters of the first-order kinetic model with rectangular distribution of floatability.

Treatment Time (s)	k (s^{-1})		R_∞ (%)		R^2	
	Cu	Ni	Cu	Ni	Cu	Ni
0	0.11967	0.09368	65.982	62.994	0.99982	0.99983
10	0.08444	0.06761	62.083	61.163	0.99981	0.99987
20	0.09912	0.07631	70.576	66.618	0.99966	0.99960
30	0.07777	0.05887	66.116	60.173	0.99994	0.99992
40	0.08081	0.05986	59.945	52.561	0.99969	0.99946

The variations of the above flotation indexes were attributed to the phase transformation during microwave treatment. Figure 18 shows the XRD patterns of the copper–nickel sulfide ore after microwave treatment for different time. After microwave treatment for 10 s, mackinawite (FeS) and cuprite (Cu_2O) were formed, with the increase in retgersite ($NiSO_4 \cdot 6H_2O$) and decrease in pentlandite due to its oxidation. After treatment time was extended to 20 s, more mackinawite was found, with less cuprite and retgersite because of further oxidation. As the treatment time increased to 40 s, all these phases decreased gradually. Moreover, rozenite ($FeSO_4 \cdot 4H_2O$) kept decreasing throughout microwave treatment, presumably following an oxidative reaction to Fe (III) sulfates [37]. Clearly, by microwave treatment, the ore began to oxidize within a short time. With the prolongation of treatment time, the oxidation would continue with the rate affected by the consumption of internal oxygen.

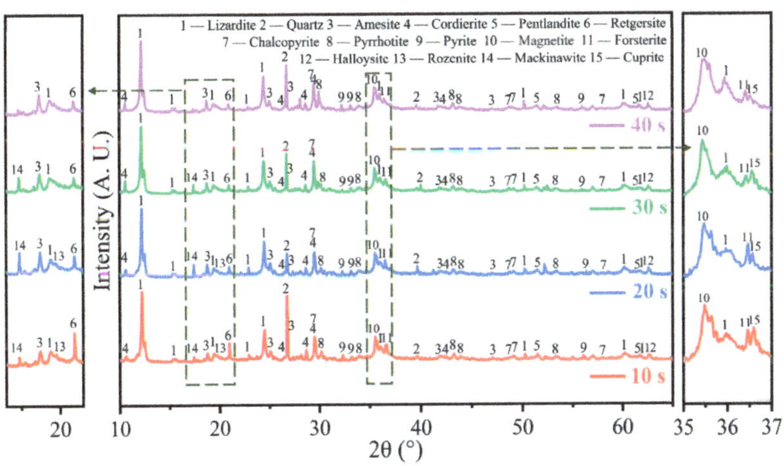

Figure 18. XRD patterns of copper–nickel sulfide ore after microwave treatment for different time.

Based on the above findings, the influence of microwave treatment on the flotation performance depended on the competition between the positive effect of the liberation of sulfide minerals and the negative effect of the oxidation of the minerals. Initially, the mineral liberation and oxidation in the ore were limited, leading to the relatively low metal recoveries after flotation. As the treatment time was extended to 20 s, the enhanced liberation and moderate oxidation of sulfide minerals under microwave irradiation contributed to the better flotation indexes. As the microwave treatment time increased continuously, the oxidation of sulfide minerals became excessive and deteriorated the flotation performance.

4. Conclusions

This study focused on the influence of microwave treatment on the grinding and flotation performance of the copper–nickel sulfide ore based on the determination of its microwave absorption capability, grinding and flotation indexes such as crack percentage, mineral liberation degree, particle size distribution, RWI, metal enrichment ratio and recovery. By measuring the electromagnetic properties, the microwave penetration depths of the ore and its main minerals were determined to confirm their different microwave absorption capabilities, which led to dissimilar microwave heating characteristics. The maximum temperature difference between the sulfide minerals and gangue minerals could reach 418 °C (chalcopyrite vs. lizardite) after microwave treatment for 20 s. The significant temperature differences between the minerals in microwave treatment provided the feasibility of microwave-assisted grinding. It was demonstrated that microwave treatment could effectively improve the ore grindability, as indicated by the increase in fine particles smaller than 0.074 mm and the decrease in RWI after grinding due to higher crack percentage and mineral liberation degree. It also affected the ore floatability because of the generation of cuprite, retgersite, and rozenite with poor floatability when the treatment time was extended. By microwave treatment for proper time, namely 20 s, the RWI of the ore decreased by 11.5%. After flotation, the enrichment ratios of Cu and Ni in the flotation concentrate increased by 0.3 and 0.2, respectively, and the recoveries of Cu and Ni increased by 4.2% and 3.1%, respectively. The findings can serve as a valuable reference for large-scale microwave processing of sulfide ores.

Author Contributions: Conceptualization, formal analysis, investigation, writing—original draft, visualization, X.F.; resources, supervision, writing—review and editing, Z.P.; formal analysis, investigation, T.Y.; visualization, methodology, M.R.; formal analysis, resources, G.L. All authors have read and agreed to the published version of the manuscript.

Funding: This work was partially supported by the Postgraduate Scientific Research Innovation Project of Hunan Province under Grant CX20220217.

Institutional Review Board Statement: Not applicable.

Informed Consent Statement: Not applicable.

Data Availability Statement: The raw data supporting the conclusions of this article will be made available by the authors on request.

Conflicts of Interest: The authors declare no conflict of interest.

References

1. Hui, S.; Li, B.; Zhou, S.; Wei, Y. Extraction of ferronickel concentrate from laterite nickel ore by reduction roasting-magnetic separation using spent cathode carbon. *Miner. Eng.* **2023**, *201*, 108194. [CrossRef]
2. Zaiman, N.F.H.N.; Shaari, N. Review on flower-like structure nickel based catalyst in fuel cell application. *J. Ind. Eng. Chem.* **2023**, *119*, 1–76. [CrossRef]
3. Golroudbary, S.R.; Kraslawski, A.; Wilson, B.P.; Lundström, M. Assessment of environmental sustainability of nickel required for mobility transition. *Front. Chem. Eng.* **2023**, *4*, 978842. [CrossRef]
4. Mudd, G.M.; Jowitti, S.M. A detailed assessment of global nickel resource trends and endowments. *Econ. Geol.* **2014**, *109*, 1813–1841. [CrossRef]
5. Liu, M.; Su, S.; Yao, Y.; Wu, X.; Cai, N.; Guan, Q. Discovery and genesis of two types of olivines and its significance to metallogeny in Jinchuan magmatic copper-nickel (PGE) sulfide deposit. *Acta Petrol. Sin.* **2020**, *36*, 1151–1170.
6. Faris, N.; Pownceby, M.I.; Bruckard, W.J.; Chen, M. The direct leaching of nickel sulfide flotation concentrates—A historic and state-of-the-art review part I: Piloted processes and commercial operations. *Miner. Process. Extr. Metall. Rev.* **2022**, *44*, 407–435. [CrossRef]
7. Forster, J.; Elliott, R.; Boucher, D.; Bobicki, E.R. High-temperature microwave properties and phase transitions of ultramafic nickel ores. *Miner. Eng.* **2021**, *172*, 107109. [CrossRef]
8. Bobicki, E.R.; Liu, Q.; Xu, Z. Microwave treatment of ultramafic nickel ores: Heating behavior, mineralogy, and comminution effects. *Minerals* **2018**, *8*, 524. [CrossRef]
9. Cai, X.; Qian, G.; Zhang, B.; Chen, Q.; Hu, C. Selective liberation of high-phosphorous oolitic hematite assisted by microwave processing and acid leaching. *Minerals* **2018**, *8*, 245. [CrossRef]
10. Yu, Q.; Ding, D.; Chen, W.; Hu, N.; Wu, L.; Zhang, Q.; Liu, Y.; Zhang, Z.; Li, F.; Xue, X.; et al. Effect of microwave pretreatment on grindability of lead-zinc ore. *Geofluids* **2021**, *2021*, 4418684. [CrossRef]
11. Toifl, M.; Hartlieb, P.; Meisels, R.; Antretter, T.; Kuchar, F. Numerical study of the influence of irradiation parameters on the microwave-induced stresses in granite. *Miner. Eng.* **2017**, *103*, 78–92. [CrossRef]

12. Adewuyi, S.O.; Ahmed, H.A.M.; Ahmed, H.M.A. Methods of ore pretreatment for comminution energy reduction. *Minerals* **2020**, *10*, 423. [CrossRef]
13. Somani, A.; Nandi, T.K.; Pal, S.K.; Majumder, A.K. Pre-treatment of rocks prior to comminution—A critical review of present practices. *Int. J. Min. Sci. Technol.* **2017**, *27*, 339–348. [CrossRef]
14. Zhang, J.; Peng, Z.; Luo, G.; Tian, R.; Rao, M. Microwave drying kinetics of chromium-rich electroplating sludge. *Metals* **2023**, *13*, 87. [CrossRef]
15. Ye, L.; Peng, Z.; Tian, R.; Tang, H.; Zhang, J.; Rao, M.; Li, G. A novel process for highly efficient separation of boron and iron from ludwigite ore based on low-temperature microwave roasting. *Powder Technol.* **2022**, *410*, 117848. [CrossRef]
16. Zhu, G.; Peng, Z.; Yang, L.; Tang, H.; Fang, X.; Rao, M. Facile preparation of thermal insulation materials by microwave sintering of ferronickel slag and fly ash cenosphere. *Ceram. Int.* **2023**, *49*, 11978–11988. [CrossRef]
17. Tang, H.; Peng, Z.; Shang, W.; Ye, L.; Luo, J.; Rao, M.; Li, G. Preparation of refractory materials from electric furnace ferronickel slag and blast furnace ferronickel slag: A comparison. *J. Environ. Chem. Eng.* **2022**, *10*, 107929. [CrossRef]
18. Lin, F.; Feng, X.; Yang, C.; Li, S.; Zhang, J.; Su, X.; Tong, T. Experimental study on improving mechanical mining efficiency of deep banded iron ore by microwave presplitting. *Int. J. Rock Mech. Min. Sci.* **2022**, *159*, 105233. [CrossRef]
19. Hartlieb, P.; Kuchar, F.; Moser, P.; Kargl, H.; Restner, U. Reaction of different rock types to low-power (3.2 kW) microwave irradiation in a multimode cavity. *Miner. Eng.* **2018**, *118*, 37–51. [CrossRef]
20. Goldbaum, M.W.; Elliott, R.; Forster, J.; Maham, Y.; Bobicki, E.R. Investigating the microwave heating behaviour of pyrrhotite tailings. *Miner. Eng.* **2020**, *146*, 106152. [CrossRef]
21. Lin, F.; Feng, X.; Lu, G.; Su, X.; Li, S.; Zhang, J. Study on microwave heating order and electromagnetic characteristics of copper and gold ores. *Rock. Mech. Rock. Eng.* **2021**, *54*, 2129–2143. [CrossRef]
22. da Silva, G.R.; Espiritu, E.R.L.; Mohammadi-Jam, S.; Waters, K.E. Surface characterization of microwave-treated chalcopyrite. *Colloids Surf. A Physicochem. Eng. Asp.* **2018**, *555*, 407–417. [CrossRef]
23. Gholami, H.; Rezai, B.; Hassanzadeh, A.; Mehdilo, A.; Yarahmadi, M. Effect of microwave pretreatment on grinding and flotation kinetics of copper complex ore. *Int. J. Miner. Metall. Mater.* **2021**, *28*, 1887–1897. [CrossRef]
24. Seiler, S.; Sánchez, G.; Pawlik, M.; Bradshaw, P.; Klein, B. Awaruite, a new large nickel resource: Flotation under weakly acidic conditions. *Minerals* **2023**, *13*, 1147. [CrossRef]
25. Li, X.; Tian, S.; Niu, Y.; Lu, B.; Ma, Y.; Xie, X.; Huang, L.; Sun, X.; Ju, S. An experimental study on microwave-assisted grinding and flotation of nickel-copper ore. *J. Kunming Univ. Sci. Technol. (Nat. Sci.)* **2022**, *47*, 7–17.
26. Liu, D.; Zhang, G.; Liu, J.; Pan, G.; Chen, Y.; Wang, M. Studies on the surface oxidation and its role in the flotation of mixed copper-nickel sulfide ore. *Powder Technol.* **2021**, *381*, 576–584. [CrossRef]
27. Kim, B.; Park, J.; Lee, J. Dielectric properites of iron sulfide in coal by microwave irradiation. *J. Korean Soc. Miner. Energy Resour. Eng.* **2013**, *50*, 838–847. [CrossRef]
28. Prameena, B.; Anbalagan, G.; Gunasekaran, S.; Ramkumaar, G.R.; Gowtham, B. Structural, optical, electron paramagnetic, thermal and dielectric characterization of chalcopyrite. *Spectrochim. Acta Part A Mol. Biomol. Spectrosc.* **2014**, *122*, 348–355. [CrossRef] [PubMed]
29. Pearce, C.I. Electrical and magnetic properties of sulfides. *Rev. Mineral. Geochem.* **2006**, *61*, 127–180. [CrossRef]
30. Forster, J.; Maham, Y.; Bobicki, E.R. Microwave heating of magnesium silicate minerals. *Powder Technol.* **2018**, *339*, 1–7. [CrossRef]
31. Ohsato, H.; Terada, M.; Kagomiya, I.; Kawamura, K.; Kakimoto, K.; Kim, E.S. Sintering conditions of cordierite for microwave/millimeterwave dielectrics. *IEEE Trans. Ultrason. Ferroelectr. Freq. Control* **2008**, *55*, 1081–1085. [CrossRef] [PubMed]
32. Lou, W.; Mao, M.; Song, K.; Xu, K.; Liu, B.; Li, W.; Yang, B.; Qi, Z.; Zhao, J.; Sun, S.; et al. Low permittivity cordierite-based microwave dielectric ceramics for 5G/6G telecommunications. *J. Eur. Ceram. Soc.* **2022**, *42*, 2820–2826. [CrossRef]
33. Peng, Z.; Hwang, J.Y.; Mouris, J.; Hutcheon, R.; Huang, X. Microwave penetration depth in materials with non-zero magnetic susceptibility. *ISIJ Int.* **2010**, *50*, 1590–1596. [CrossRef]
34. Vinnett, L.; Waters, K.E. Representation of kinetics models in batch flotation as distributed first-order reactions. *Minerals* **2020**, *10*, 913. [CrossRef]
35. Yu, D.; Utigard, T.A. TG/DTA study on the oxidation of nickel concentrate. *Thermochim. Acta* **2012**, *533*, 56–65. [CrossRef]
36. Feng, B.; Feng, Q.; Lu, Y. A novel method to limit the detrimental effect of serpentine on the flotation of pentlandite. *Int. J. Miner. Process* **2012**, *114*, 11–13. [CrossRef]
37. Xu, L.; Xie, Q.; Chen, T.; Li, P.; Yang, Y.; Zhou, Y. Constraint of nanometer-sized pyrite crystals on oxidation kinetics and weathering products. *J. Nanosci. Nanotechnol.* **2017**, *17*, 6962–6966. [CrossRef]

metals

Article

A Pretreatment of Refractory Gold Ores Containing Sulfide Minerals to Improve Gold Leaching by Ammonium Thiosulfate: A Model Experiment Using Gold Powder and Arsenic-Bearing Sulfide Minerals

Takunda Joseph Mhandu [1,*], Ilhwan Park [2,*], Sanghee Jeon [3], Sohta Hamatsu [1], Yogarajah Elakneswaran [2], Mayumi Ito [2] and Naoki Hiroyoshi [2]

[1] Division of Sustainable Resource Engineering, Graduate School of Engineering, Hokkaido University, Sapporo 060-8628, Japan; sohta1415@gmail.com

[2] Division of Sustainable Resource Engineering, Faculty of Engineering, Hokkaido University, Sapporo 060-8628, Japan; elakneswaran@eng.hokudai.ac.jp (Y.E.); itomayu@eng.hokudai.ac.jp (M.I.); hiroyosi@eng.hokudai.ac.jp (N.H.)

[3] Department of Earth Resources Engineering and Environmental Science, Faculty of International Resources Science, Akita University, Akita 010-0852, Japan; jeon@gipc.akita-u.ac.jp

* Correspondence: takundajoseph.mhandu.i9@elms.hokudai.ac.jp (T.J.M.); i-park@eng.hokudai.ac.jp (I.P.)

Abstract: The use of thiosulfate to extract gold from refractory ores is promising because of its non-toxicity and high selectivity. Sulfide minerals (i.e., pyrite, arsenopyrite, chalcopyrite), major gold carriers in refractory gold ores, however, hinder gold extraction due to the high consumption of a lixiviant. In this study, a new method to improve gold extraction from sulfide bearing gold ores is proposed based on the model experiments using a mixture of gold powder and arsenopyrite-bearing sulfide (HAsBS) ore. The effects of HAsBS ore on gold leaching in ammonium thiosulfate solutions were investigated, and it was found that gold extraction in the presence of HAsBS ore was suppressed because of the unwanted decomposition of thiosulfate on the surface of sulfide minerals. To improve gold extraction in the presence of the sulfide minerals, this study investigated the effects of the pretreatment of HAsBS ore using ammonium solutions containing cupric ions and confirmed that HAsBS ore was oxidized in the pretreatment and its surface was covered by the oxidation products. As a result, thiosulfate consumption was minimized in the subsequent gold leaching step using ammonium thiosulfate, resulting in an improvement in gold extraction from 10% to 79%.

Keywords: gold; arsenopyrite; ammonium thiosulfate; cyanide; leaching; refractory ores

1. Introduction

Gold (Au) is an important metal in the livelihood of humans and is currently used for jewelry (53%), financial purposes (i.e., investment and central banks) (39%), and electric and electronic equipment (6%), and the remaining 2% is used for other applications such as medical and pharmaceutical products [1–3]. This demand for gold can be sustained by discovering new high-grade gold deposits, but finding the high-grade deposits has become difficult. Wilson et al. (2022) reported that there has been a decrease in the gold grades in global gold deposits from 10 g/t in the 1970s [4] to less than 1.5 g/t [5] currently, of which the deposits being discovered have more unwanted materials that are difficult to process. Hence, the gold mining industry has been driven by the demand for gold to find ways to process lower grade and refractory gold deposits over the past 50 years [6,7]. In these refractory ores, gold commonly occurs as free particles of gold and electrum, with the inclusion of these phases in the sulfide mineral matrix [8]. The encapsulation of gold in sulfide matrices as well as its association with other minerals (e.g., arsenopyrite (FeAsS), chalcopyrite ($CuFeS_2$), and pyrite (FeS_2)) renders these gold ores refractory. For the past century, cyanide has been the dominant lixiviant for the extraction of gold from its hypergenic ores, but due to its toxicity, environmental protection organizations and governments are pushing for the ban of cyanide usage in gold extraction processes [9–12]. To address this problem, alternative lixiviants such as halides, thiourea, and thiosulfate have been suggested by many researchers. Among them, thiosulfate is the most promising lixiviant due to its low toxicity, high selectivity to gold, and suitability for use with refractory gold ore containing carbonaceous materials [13–16]. The industrial application of thiosulfate is, however, still limited due to unacceptable high thiosulfate consumption and the passivation of gold, though it is notable that Goldstrike Mine, Nevada, USA, is using thiosulfate [17,18]. Decomposition of the thiosulfate, as well as the passivation of gold, further accelerates when sulfide minerals (e.g., pyrite and/or arsenopyrite) exist in the refractory ore leaching system [19–21]. Moreover, in the case of high arsenopyrite-bearing sulfide ore (HAsBS), it causes acid mine drainage containing arsenic (As) [22], which is harmful to micro- and macro-organisms. In ammonium thiosulfate leaching, HAsBS causes the decomposition of thiosulfate and the formation of a passivation layer (i.e., cuprous sulfide (Cu_2S), sulfur (S)) on the gold surface, which hinders the dissolution of gold by acting as a barrier between thiosulfate ions and gold grains [21]. To enhance the gold extraction in the ammoniacal thiosulfate leaching system, roasting has been the predominantly used technique to oxidize refractory sulfide gold ores before leaching [23]. By roasting, dense sulfide minerals are oxidized and converted to porous oxides, which increases the contact between gold and the lixiviant, and minimizes thiosulfate decomposition, resulting in the enhancement of gold extraction. In the case of HAsBS, however, roasting generates toxic gases containing sulfur oxide gases (SO_x) and arsenic oxides (As_2O_3 or As_2O_5) making it an unsuitable technique to treat HAsBS [24,25]. These toxic emissions can lead to leaks of arsenic and potential incidents of exposure, highlighting the unsuitability of roasting for HAsBS ores [26,27].

To avoid toxic arsenic containing gas formation, arsenic-bearing sulfide minerals in refractory gold ores can be pre-oxidized in low-temperature processes to minimize thiosulfate decomposition in the same manner as high-temperature roasting. Pre-oxidation of sulfide minerals using ammonium solutions containing cupric ions can be considered as a pretreatment process for refractory gold ores at ambient temperature and pressure conditions. In this pretreatment process, sulfide minerals are oxidized by cupric ions (or cupric amine complexes), and associated metal species such as copper in the sulfide minerals are extracted into aqueous phase by forming soluble complexes with ammonium, while iron, the dominant metal in the sulfide minerals, is precipitated as metal oxyhydroxides on the surface of sulfide grain. This may minimize thiosulfate decomposition due to sulfide minerals in the subsequent gold extraction process using ammonium thiosulfate.

This paper proposes the pre-oxidation of sulfides using ammonium solutions containing cupric ions as a low-temperature pretreatment suitable for arsenic-bearing refractory ore for improving gold extraction in the succeeding ammonium thiosulfate leaching. To demonstrate the effects of the proposed pre-oxidation, model experiments were conducted using the mixture of gold powder and HAsBS ore. The effects of the HAsBS ore on gold extraction were investigated to confirm the suppression mechanism of gold extraction with the ore. The effects of the pretreatment using ammonium solutions containing cupric ions on the subsequent gold leaching using ammonium thiosulfate was demonstrated.

2. Materials and Methods

2.1. Materials

The HAsBS sample was supplied by Kitabita Mine, Yamaguchi Prefecture, Japan. Table 1 shows the chemical composition of the HAsBS sample determined using the X-ray fluorescence spectrometer (XRF, EDXL300, Rigaku Corporation, Tokyo, Japan). Quantitative X-ray diffraction (XRD, Multiplex, Rigaku Corporation, Tokyo, Japan) analysis result (Figure 1) showed that the HAsBS sample was composed of arsenopyrite, chalcopyrite, pyrite, pyrrhotite, and quartz.

Table 1. Chemical composition of arsenopyrite sample.

Elements	Fe	As	S	O	Mg	Si	Cu	Br	Zn	Sb	Others
wt.%	29.2	25.1	15.4	14.5	11.5	3.9	0.24	0.1	0.03	0.01	0.02

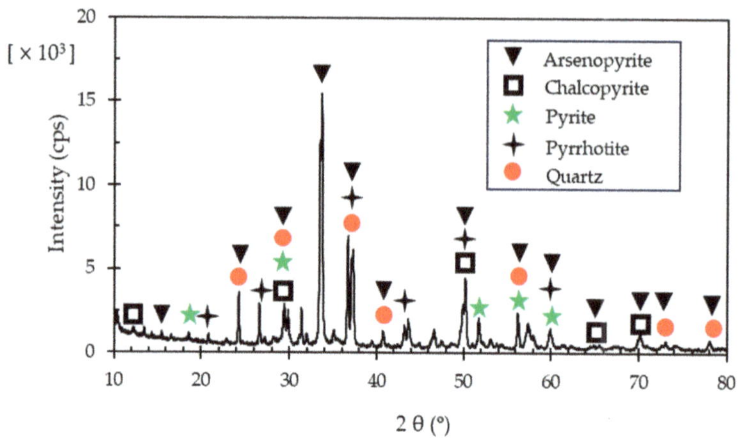

Figure 1. The X-ray powder diffraction (XRD) patterns of HAsBS ore.

The ore sample was crushed using a jaw crusher (1023-A, Yoshida Manufacturing Co., Ltd., Sapporo, Japan) with an aperture size of 1 mm, and ground using a Retsch RS 200-disc mill (Retsch Inc., Hean, Germany) to sizes of 80% less than 38 μm. Leaching experiments were conducted using gold powder (99.99%), sodium thiosulfate pentahydrate ($Na_2S_2O_3 \cdot 5H_2O$), ammonium sulfate ($(NH_4)_2SO_4$), ammonia water ($NH_4(OH)$, 31%), and cupric sulfate ($CuSO_4$ $5H_2O$), which were all obtained from Wako Pure Chemical Industries, Ltd., Japan. To characterize the surface of gold during leaching, a gold foil (thickness 0.127 mm and 99.99% purity) obtained from Sigma-Aldrich, Corporation, St. Louis, MO, USA, was used.

2.2. Leaching Experiments

Batch leaching experiments were carried out in 50 mL Erlenmeyer flasks containing a known amount of HAsBS (1 g) and 10 mg of gold powder in 10 mL of ammonium thiosulfate solution consisting of 1 M $Na_2S_2O_3 \cdot 5H_2O$, 0.25 M $(NH_4)_2SO_4$, 0.5 M NH_4OH, and 10 mM $CuSO_4 \cdot 5H_2O$. The flasks were shaken in a thermostat water bath shaker at 25 °C for 24 h with a constant shaking amplitude and frequency of 40 mm and 120 min^{-1}, respectively. The effects of leaching time (2–24 h), solid-to-liquid ratio (1–20%), $CuSO_4$ concentration (0.1–15 mM), $Na_2S_2O_3$ concentration (0.1–2 M) on gold dissolution were investigated in the leaching experiments. After leaching, the suspensions were filtered using 0.2 μm membrane filters (Sartorius AG, Göttingen, Germany), and the filtrates were analyzed by inductively coupled plasma atomic emission spectroscopy (ICP–AES 9820, Shimazu Corporation Japan (margin of error ±2%)) to measure the concentration of dissolved elements and by ultraviolet-visible spectroscopy (UV–Vis, UV–970, Japan Spectroscopic Co., Ltd., Tokyo, Japan) to analyze the thiosulfate and sulfur species. The residues were washed thoroughly with deionized (DI) water, dried in a vacuum drying oven at 40 °C, and analyzed by XRF, X-ray photoelectron spectroscopy (XPS, JPS–9200, JEOL Ltd., Akishima, Japan), and scanning electron microscopy with energy dispersive X-ray spectroscopy (SEM–EDX, JSM–IT200, JEOL Ltd., Tokyo, Japan) for chemical and surface characterization, respectively.

To determine the amount of dissolved metal in the leaching experiments of gold powder in ammonium thiosulfate solutions, the metal extraction was calculated using the following equation.

$$\text{Metal extraction} = \frac{C_m \times V}{m \times M} \times 100\%, \quad (1)$$

where C_m is the concentration of dissolved metal in mg/L, V is the volume of leaching solution in L, m is the mass of the sample (g), and M the metal content in the sample (mg/g), respectively. In addition, thermodynamic calculations using Geochemist's Workbench (GWB) software were carried out for the Fe–As–S–H_2O and Fe–S–H_2O systems to predict the dissolution behavior of the metal species in the HAsBS ore for the conditions used in the leaching experiments.

2.3. Characterization of the Passivation Layer on Gold Surface

Batch leaching experiments were conducted to characterize the passivation layer formed on the gold grains in ammonium thiosulfate solutions. Leaching was conducted using 4 mm × 4 mm gold foil, 1 g HAsBS, and 10 mL of ammonium thiosulfate solution, all of which were put in 50 mL flask and shaken (i.e., a thermostat water bath shaker at 25 °C; amplitude 40 mm; shaking speed 120 min^{-1}) for 24 h. The residues were washed thoroughly with DI water, dried in a vacuum drying oven at 40 °C, and analyzed by XRF, XPS, and SEM–EDX for chemical and surface characterization.

2.4. Pretreatment Experiments Using Ammonium Solution Containing Cupric Ions

Pre-oxidation experiments were carried out by mixing 10 mg gold powder, 1 g HAsBS, and 10 mL of solution containing 0.5 M NH_3, 0.5 M NH_4Cl, and 10 mM $CuCl_2$ in a 50 mL Erlenmeyer flask and shaking in a thermostat water bath shaker at 25 °C (amplitude 40 mm; shaking speed 120 min^{-1}) for 24 h. After 24 h, shaking was stopped for decantation and 5 mL of supernatant was taken for analysis. Subsequent gold leaching was carried out by adding 5 mL of 2 M $Na_2S_2O_3$ solution containing 0.5 M NH_3, 0.5 M NH_4Cl, and 10 mM $CuCl_2$ to the flask and shaking the flask for another 24 h. After gold leaching, solutions and solid residues were collected by filtration, and analyzed by ICP-AES and SEM–EDX, respectively. Control experiments (gold leaching experiments without pre-oxidation) were carried out by shaking 10 mg gold powder, 1 g HAsBS, and 10 mL of solution containing 1 M $Na_2S_2O_3$, 0.5 M NH_3, 0.5 M NH_4Cl, and 10 mM $CuCl_2$ in 50 mL Erlenmeyer flask at 25 °C for 24 h.

To evaluate the effect of pretreatment on thiosulfate decomposition, leaching experiments were carried out by mixing 4 g of HAsBS (i.e., pretreated, and untreated) with 40 mL of thiosulfate solution (1 M $Na_2S_2O_3$, 0.5 M NH_3, 0.5 M NH_4Cl) with and without 10 mM $CuCl_2$ in a 50 mL Erlenmeyer flask using magnetic stirring. A 2 mL suspension was collected at regular time intervals (20 min) and the filtrate was analyzed by UV–Vis.

3. Results and Discussion

3.1. Effect of HAsBS Ore on Gold Dissolution

The suppression mechanism of gold extraction from refractory ores is classified into two phenomena: one is the physical encapsulation of gold grains in the sulfide matrix, which prevents the contact of the lixiviant with the gold grains. The second is a chemical suppression mechanism that includes thiosulfate decomposition and passivation of the gold grains. The suppressive effect of physical encapsulation can be overcome by ultra-fine grinding to expose the gold grain to the lixiviant, but chemical suppression remains even after ultra-fine grinding is conducted. As a first step in the present study, the chemical suppressive effects of sulfide minerals on gold extraction were evaluated by leaching experiments using the mixture of gold powder and HAsBS ore (Figure 2). In the absence of HAsBS, over 90% of gold was extracted within 8 h, and more than 99% gold extraction was attained after 24 h. This fast-leaching kinetics matches with previous studies [28–30].

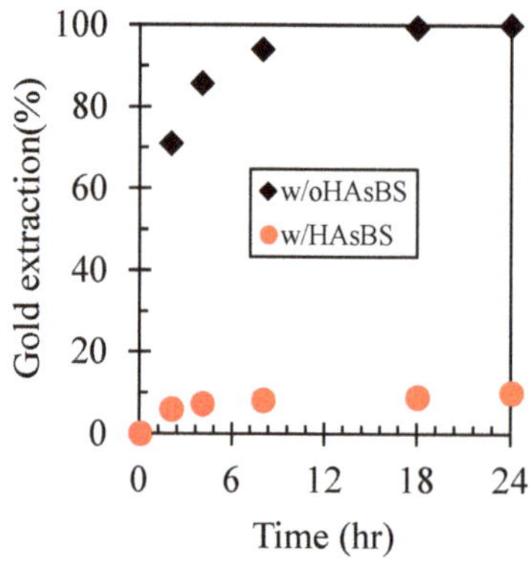

Figure 2. The effects of HAsBS ores on gold extraction in ammonium thiosulfate solutions.

The dissolution reaction of gold in ammonium thiosulfate has been reported by many authors and is summarized as: (Equation (2)) [27,29–34].

$$Au + 2S_2O_3^{2-} + Cu(NH_3)_4^{2+} \rightarrow Au(S_2O_3)_2^{3-} + Cu(NH_3)_4^{+} \tag{2}$$

In Equation (2), gold is oxidized by the cupric ammine complex ($Cu(NH_3)_4^{2+}$), and the extracted gold (Au^+) is stabilized by forming a complex with thiosulfate anions (i.e., $Au(S_2O_3)_2^{3-}$). The cuprous ammine complex ($Cu(NH_3)_4^+$) formed in Equation (2) is then oxidized to a cupric ammine complex ($Cu(NH_3)_4^{2+}$) with dissolved oxygen in Equation (3). The overall reaction of gold dissolution from Equations (2) and (3) is summarized by Equation (4).

$$4Cu(NH_3)_4^{+} + O_2 + 2H_2O \rightarrow 4Cu(NH_3)_4^{2+} + 4\,OH^- \tag{3}$$

$$4Au + 8S_2O_3^{2-} + O_2 + 2H_2O \rightarrow 4Au(S_2O_3)_2^{3-} + 4\,OH^- \tag{4}$$

In the overall reaction of gold dissolution (Equation (4)), copper ammines ($Cu(NH_3)_4^{2+}/(Cu(NH_3)_4^{+})$) do not appear, and they are considered to be a catalyst for gold extraction. In the solution phase, the decomposition of thiosulfate occurs; that is, thiosulfate ions are oxidized by the cupric ammine complex to form tetrathionate ($S_4O_6^{2-}$) as follows [31,32]:

$$2Cu(NH_3)_4^{2+} + 2S_2O_3^{2-} \rightarrow 2Cu(NH_3)_4^{+} + S_4O_6^{2-} \tag{5}$$

Because of this reaction (Equation (5)), thiosulfate ions, needed to extract gold, are decomposed, but as Figure 2 shows, more than 99% of gold was extracted in the absence of the HAsBS ore; this implies that the effects of thiosulfate decomposition can be considered negligible.

In the presence of HAsBS, a maximum of 10% gold extraction was attained after 24 h, which was much lower compared to the absence of HAsBS where over 99% extraction was obtained. There are two possible reasons that can explain the drastic reduction in gold extraction in the presence of HAsBS, which are: (1) the formation of a passivation layer on the surface of gold that retarded the dissolution of gold [21], and (2) the accelerated decomposition of thiosulfate by associated sulfide minerals [19]. In the following sections, the above hypotheses are discussed based on the experimental results.

3.2. Oxidation of HAsBS Ore in Ammonium Thiosulfate Solution

During gold leaching in ammonium thiosulfate solutions, arsenopyrite and pyrite, the major sulfide minerals in HAsBS ore (Figure 1), are oxidized by the cupric ammine complex. Thermodynamic calculations performed using GWB software for the Fe–As–S–H_2O and

Fe–S–H_2O systems (Figure 3) showed that iron arsenate ($FeAsO_4$) and hematite (Fe_2O_3) are the stable phase for the iron and arsenic species at the observed pH (9.8) and Eh (0.15 V) conditions in the leaching experiments, shown as the cross point of the red dotted lines in Figure 3. Based on this, the possible reactions for arsenopyrite and pyrite oxidation are assumed to be:

$$FeAsS + 14Cu(NH_3)_4{}^{2+} + 8H_2O \rightarrow FeAsO_4 + SO_4{}^{2-} + 14Cu(NH_3)_4{}^{+} + 16H^{+} \quad (6)$$

$$2FeS_2 + 12Cu(NH_3)_4{}^{2+} + 7H_2O \rightarrow Fe_2O_3 + 4SO_4{}^{2-} + 12Cu(NH_3)_4{}^{+} + 14H^{+} \quad (7)$$

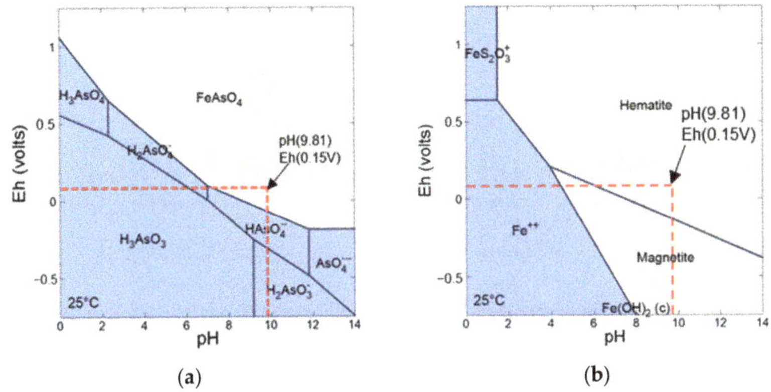

Figure 3. Eh–pH diagram of the (**a**) Fe–As–S–H_2O system, (**b**) Fe–S–H_2O system. (Concentrations of Fe, As, and S were taken to be 0.001 mg/L, 0.001 mg/L, and 0.1 mg/L.)

To confirm the proposed oxidation reactions for arsenopyrite and pyrite, soluble Fe, As, and S concentrations during the leaching experiments with HAsBS were analyzed (Figure 3). The dissolved sulfur concentration increased from 39 to 47 g/L in the first 6 h of leaching, Figure 4b. This is consistent with reaction (6) and (7), where sulfate ions are released by the oxidation of HAsBS. After 6 h, the soluble sulfur concentration started to decrease, and the concentration reached 36 g/L at 24 h. The decline in soluble sulfur concentration indicates that precipitation of the sulfur species occurs during the leaching. This may be interpreted by assuming the decomposition of thiosulfate due to the presence of an excess amount of dissolved oxygen. According to Rodriguez et al. [33], an excess amount of dissolved oxygen favors the oxidation of thiosulfate and its complexation with Cu; oxygen causes the oxidation of Cu(I) to Cu(II), at the same time decomposing thiosulfate to form $(S_2O_3)_3{}^{5-}$, sulfite($SO_3{}^{2-}$), sulfate($SO_4{}^{2-}$), and elemental sulfur(S) as shown in Equations (8) and (9) [34]. In addition, the rate of oxidation of Cu(I) to Cu(II) is slower than the reduction of Cu(II) by thiosulfate, which may result in the formation of Cu(I)-sulfide precipitates, thus causing a reduction in the sulfur concentration.

$$3S_2O_3{}^{2-} + 6H_2O \rightleftharpoons 2SO_4^{2-} + 4S^0 + 2OH^- \quad (8)$$

$$3S_2O_3{}^{2-} + 6OH^- \rightleftharpoons 4SO_3{}^{2-} + 2S^{2-} + 3H_2O \quad (9)$$

$$S_2O_3{}^{2-} \rightarrow SO_3{}^{2-} + S^0 \quad (10)$$

As shown in Figure 4a, the extraction of Fe and As was very low with about 0% and less than 0.4% after 24 h of leaching, respectively. This indicates that both Fe and As species are not stable in the aqueous phase under the leaching conditions, and they are precipitated, probably, as ferric arsenate, hematite, or ferric hydroxides. Such precipitates might affect the gold extraction; they may cover the surface of gold powder, suppressing gold extraction, and this is discussed in detail in the next section, Section 3.3.

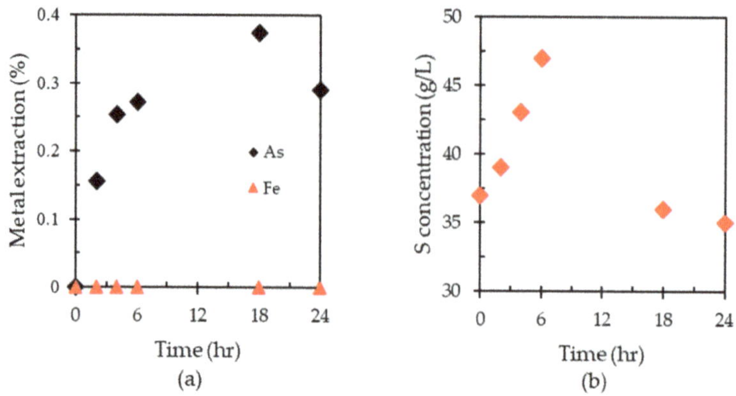

Figure 4. The dissolution behavior of HAsBS during gold leaching: (**a**) extraction of Fe and As, and (**b**) change in S concentration with time.

3.3. *Effects of the Passivation Layer on Gold Extraction*

As highlighted in Section 3.2, the second possible reason that can cause the reduction in gold extraction in the presence of HAsBS is the formation of a passivation layer. A passivation layer covers the surface of the gold and prevents interaction between the gold and the lixiviant, resulting in low gold extraction. In a study by Yang et al. (2015), characterization of the passivation layer after ammonium thiosulfate leaching in the presence of arsenopyrite revealed that the passivation layer was comprised of cuprous sulfide (Cu_2S), elemental sulfur (S), iron oxyhydroxide (FeOOH), and iron arsenate ($FeAsO_4$) species on the surface of gold. To confirm the presence of these materials, this study analyzed the surface of the leached gold foils using SEM–EDX and XPS. Figure 5 shows the results of SEM–EDX analysis for gold foil before and after leaching with HAsBS ores.

The SEM–EDX images (Figure 5) and EDX spectrum (Figure 5(a-1,b-1)) did not detect any sulfur, arsenic, or copper species but showed the presence of oxygen and iron on the surface of gold after leaching. This may be because the amounts of the passivating materials on the gold surface were below the detection limit of SEM–EDS analysis. To detect a small amount of passivating materials, XPS was used to characterize the gold surface in the presence (i) and absence (ii) of HAsBS, (Figure 6).

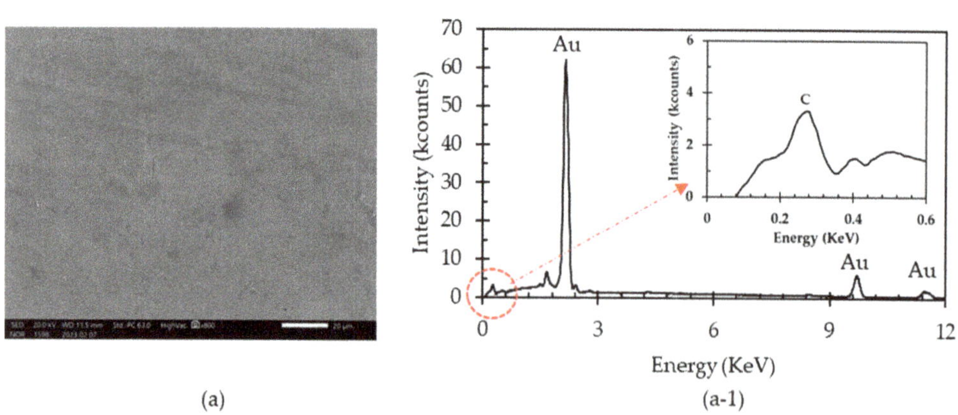

Figure 5. *Cont.*

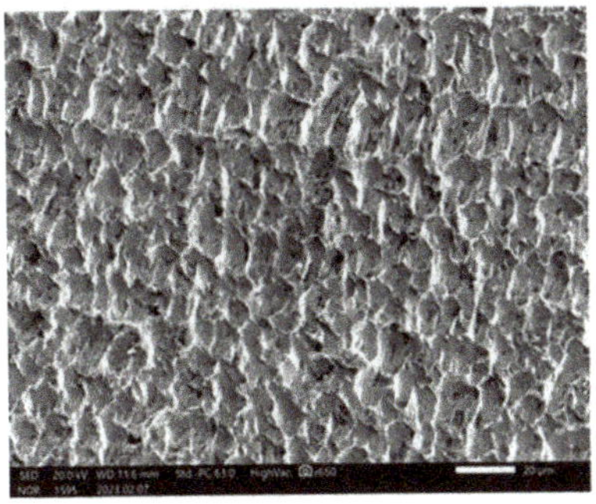

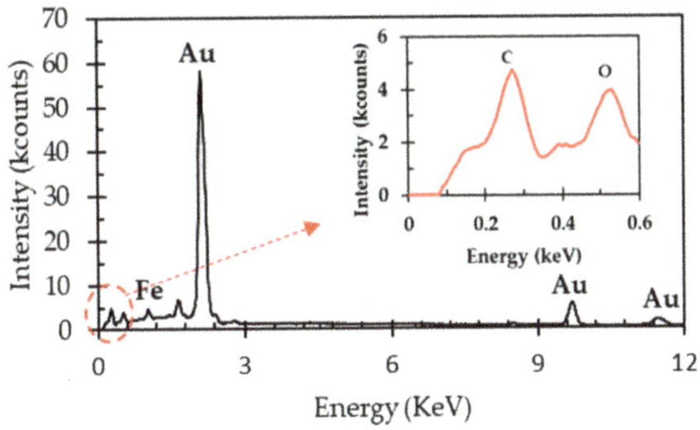

(b) (b-1)

Figure 5. SEM–EDX photomicrographs of gold foil (**a**) before leaching and the corresponding EDX spectrum (**a-1**), (**b**) gold foil after leaching and the subsequent EDX spectrum (**b-1**).

As shown in Figure 6, in the XPS spectrum, signals from Cu, Fe, As, S, and O were detected on the surface of the gold foil leached in the presence of HAsBS (i). In the deconvolution results for the Cu 2p3/2 spectrum, Figure 6a, a strong peak at 932.5 eV was assigned to Cu(I) bound to a sulfur species as Cu_2S [35], and minor peaks at 934.5 eV and 936.0 eV corresponding to Cu(II)-oxide and Cu(II)-sulfate were detected, respectively [36]. In the deconvoluted spectrum of S 2p3/2, peaks were detected at 163.0 eV and 161.7 eV and were assigned to disulfide and monosulfide sulfur species combined with copper species, respectively [37], and at 165.5 eV corresponding to elemental sulfur. In addition, satellite peaks were observed at 167.3 eV and 167.9 eV, which could be S associated with the sulfate species [38]. In the case of oxygen, the O 1s deconvoluted peaks at 531.9 eV and 534.3 eV were observed and were all attributed to oxygen in hydroxyl ions of Fe(III) hydroxide [39], while another small peak at 536.0 eV was assigned to the oxygen of water molecules. The iron spectrum showed peaks at 706.8 eV and 710.0 eV corresponding to Fe(II)-AsS and Fe(III)-AsS in the bulk FeAsS, respectively. The other iron peaks detected were at 711.7 eV corresponding to Fe(III)–hydroxide and at 713.7 eV attributed to similar Fe(III)-SO_4 species [40]. In addition, there were satellite iron peaks observed at 715.8 eV and 715.1 eV, which were assigned to hydrated Fe(III) oxide [41]. The arsenic spectra exhibited strong peaks at 41.3 eV and 45.5 eV, assigned to arsenic in the bulk FeAsS and As(III)-oxide, respectively. Furthermore, a peak at 46.2 eV was detected and assigned to (As(V)-O) species. The results from the XPS validated the presence of a passivation layer for the gold foil leached in the presence of HAsBS ore and were consistent with the work conducted by Yang et al. (2015) but did not give a clear indication of the source of elemental sulfur that formed on the surface of the gold foil.

To determine the source of the sulfur species observed on the gold foil, XPS characterization was undertaken for the gold foil leached in the absence of the HAsBS (ii) and the results are outlined in Figure 6(aii–eii). The XPS spectrum of Cu 2p3/2 (Figure 6(aii)) showed a peak centered at 933.5 eV assigned to Cu(II) oxide, while peaks observed at 931.9 eV and 935.6 3 eV were assigned to Cu(II) sulfide and hydrated Cu(II) sulfate, respectively [36]. For the sulfur spectrum (Figure 6(bii)), two peaks at 161.9 eV and 164.1 eV were shown and assigned to S^{2-} and elemental sulfur (S), respectively. Oxygen peaks at 533.9 eV and 531.3 eV were assigned to the hydroxide ion (OH^-), and peaks at 532.4 eV and 536.2 eV were attributed to the oxygen of water and Cu(II) species, respectively [42]. These results show that even in the absence of HAsBS, elemental sulfur was formed on the gold surface. This indicates that the major source of sulfur in the passivation layer may be formed by the decomposition of thiosulfate.

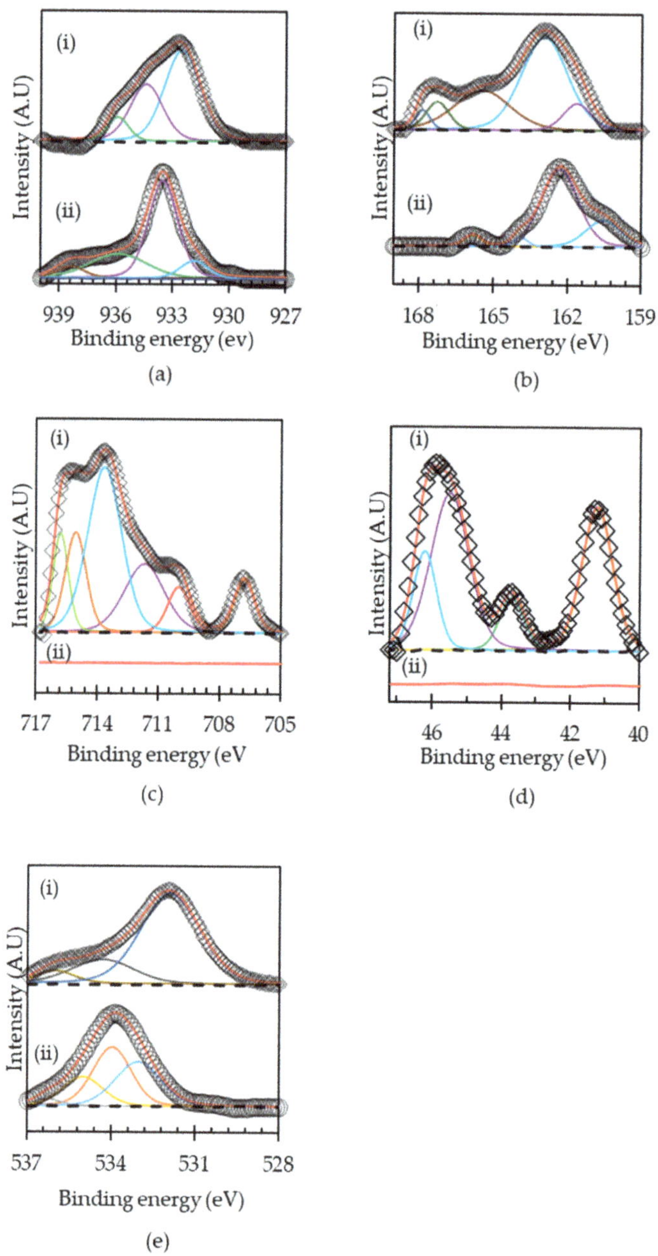

Figure 6. XPS spectra of gold foil residue leached in the presence of HAsBS (**i**) and without HAsBS (**ii**): (**a**) Cu 2p, (**b**) S 2p, (**c**) Fe 2p (**d**) As 3d, (**e**) O 1 s. The red lines denote the fitting line for the XPS spectra, whilst the green, yellow, purple, and blue lines represent the deconvoluted spectra for the respective elements.

The XPS analysis established that during gold leaching, the major passivating species on the gold surface are elemental sulfur, Cu_2S/Cu_2O, and Fe(III)−O. The following Table 2 gives a summary of the assignments used to interpret the XPS spectra for the surface of the gold in the presence and absence of HAsBS.

Table 2. XPS peak parameters for Fe 2p, As 3d, S 2p, O1 s, and Cu 2p spectra.

Spectral Peak	Binding Energy	FWHM	Line Shape	Chemical Species	Content (at. %)	References
Fe $2p_{3/2}$ [a]	706.8	1.2	GL (20)	Fe(II)–AsS	7.8	[22]
Fe $2p_{3/2}$ [a]	710.0	2.4	GL (20)	Fe(III)–AsS	7.3	[40]
Fe $2p_{3/2}$ [a]	711.7	0.9	GL (20)	Fe(III)–O	19.9	[43,44]
Fe $2p_{3/2}$ [a]	713.7	1.3	GL (20)	Fe(III)–SO_4^{2-}	39.5	[45,46]
As $3d_{5/2}$ [a]	41.3	1.1	GL (20)	As(O)	31.6	[43]
As $3d_{5/2}$ [a]	43.7	0.9	GL (20)	As(I)–O	9.5	[43]
As $3d_{5/2}$ [a]	45.5	1.3	GL (20)	As(III)–O	42.6	[47]
As $3d_{5/2}$ [a]	46.2	0.8	GL (20)	As(V)–O	16.4	[47]
S $2p_{3/2}$ [a]	161.7	1.3	GL (20)	Monosulfide (S^{2-})	9.3	[43]
S $2p_{3/2}$ [a]	163.0	2.2	GL (20)	Disulfide (S_2^{2-})	52.7	[43]
S $2p_{3/2}$ [a]	165.5	2.7	GL (20)	Elemental (S)	27.0	[43]
S $2p_{3/2}$ [a]	167.9	0.7	GL (20)	Thiosulfate (SO_3^{2-})	3.5	[43]
S $2p_{3/2}$ [a]	167.3	1.1	GL (20)	Thiosulfate (SO_3^{2-})	7.6	[43]
S $2p_{3/2}$ [b]	160.6	1.7	GL (20)	Monosulfide (S^{2-})	22.8	[43]
S $2p_{3/2}$ [b]	162.3	1.77	GL (20)	Disulfide (S_2^{2-})	69.6	[43]
S $2p_{3/2}$ [b]	163.9	0.68	GL (20)	Elemental sulfur (S^0)	3.0	[45]
S $2p_{3/2}$ [b]	165.9	0.81	GL (20)	Sulfate (SO_4^{2-})	4.52	[43]
Cu $2p_{3/2}$ [a]	932.5	2.3	GL (20)	Cu(I)–S	57.9	[48]
Cu $2p_{3/2}$ [a]	934.5	2.1	GL (20)	Cu(II)–O	33.5	[35,49]
Cu $2p_{3/2}$ [a]	936.0	1.3	GL (20)	Cu(II)–sulfate.	8.8	[36]
Cu $2p_{3/2}$ [b]	931.9	1.9	GL (20)	Cu(I)–S	57.5	[48]
Cu $2p_{3/2}$ [b]	933.5	3.6	GL (20)	Cu(II)–O/$Cu(OH)_2$	26.8	[50,51]
Cu $2p_{3/2}$ [b]	935.9	1.7	GL (20)	$Cu(OH)_2$	6.1	[21,51]
Cu $2p_{3/2}$ [b]	938.4	1.7	GL (20)	Cu(II)–sulfate.	9.5	[36]
O 1s [a]	531.9	2.5	GL (20)	Hydroxyl oxygen (OH^-)	73.9	[22]
O 1s [a]	534.3	1.6	GL (20)	(O)attached to water	6.5	[38]
O 1s [a]	536.0	2.6	GL (20)	(O)attached to water	19.6	[38]
O 1s [b]	533.1	1.9	GL (20	Hydroxyl oxygen (OH^-)	36.6	[51]
O 1s [b]	534	1.6	GL (20)	(O)attached to water	39.7	[51]
O 1s [b]	535.0	1.7	GL (20)	(O)attached to water	20.1	[51]
O 1s [b]	536.3	1.1	GL (20)	(O)attached to water	3.5	[51]

Keys: a—gold foil leached with HAsBS, b—gold foil leached without HAsBS.

Using the data in Table 2, the presence of a passivation layer composed of Cu_2S, S, and FeOOH was confirmed on the gold surface in the presence of HAsBS by XPS analysis. The formation of passivation materials (Cu_2S and S) was, however, also observed even without HAsBS, indicating that the major source of sulfur in the passivation materials comes from the decomposition of thiosulfate. Considering these results, the effects of the passivation layer were limited and may not cause low gold extraction in the presence of HAsBS ore as used in this study but could be caused by thiosulfate decomposition.

3.4. *Effect of HAsBS on Thiosulfate Consumption*

In the previous section, the formation of a passivation layer on gold with HAsBS and its effects were discussed. Another possible mechanism for the low gold extraction is thiosulfate decomposition induced by sulfide minerals. Therefore, the stability of thiosulfate in the presence of HAsBS was evaluated using UV–Vis spectral analysis to confirm thiosulfate decomposition by HAsBS. The UV–Vis spectral analyses for 0.5 M NH_4Cl/0.5 M NH_3 buffer solutions containing 100 μM of sodium thiosulfate or sodium sulfate are shown in Figure 7. Thiosulfate gave a strong absorbance peak at 214–216 nm, which is in line with previous reports [52,53]. The absorbance for sulfate ions was very low in the measured range (190–290 nm), and thus it can be assumed that absorbance at 215 nm (Abs_{215}) corresponds to thiosulfate.

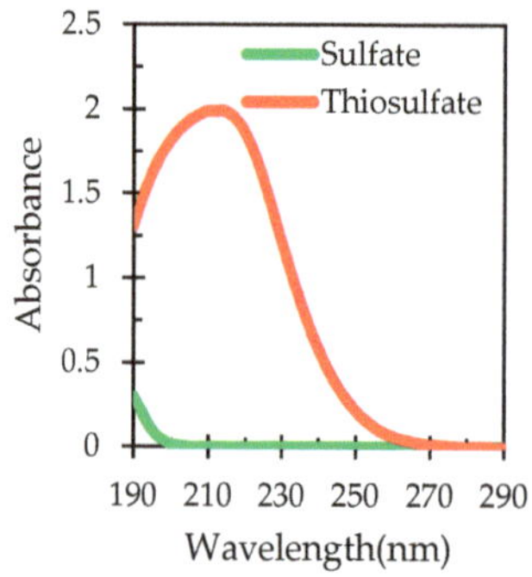

Figure 7. UV–Vis spectra of sulfate (SO_4^{2-}) and thiosulfate ($S_2O_3^{2-}$).

The stability of thiosulfate in 0.5 M NH_4Cl/0.5 M NH_3 buffer solutions containing 100 μM of sodium thiosulfate was assessed with and without HAsBS (Figure 8). The effects of HAsBS addition with time on the absorbance at 215 nm (Abs_{215}), corresponding to thiosulfate, were measured (Figure 9). The vertical axis of Figure 9 shows Abs_{215} at t min normalized with initial, at t = 0 min, $Abs_{215}(0)$, given by the ratio $Abs_{215}(t)/Abs_{215}(0)$. Without HAsBS, the normalized absorbance slightly decreased with time, indicating that thiosulfate tends to self-decompose [17]. Even with the self-decomposition, the normalized absorbance remained over 0.60 after 80 min, suggesting that about 65% of thiosulfate remains in the experiments. The decreasing rate of the normalized absorbance was strongly affected by the addition of HAsBS. With HAsBS, the normalized absorbance rapidly decreased to 0.20 at 20 min, and it reached 0.05 at 80 min. This suggests that thiosulfate decomposition was accelerated by HAsBS. Xu et al. (2017) reported that thiosulfate decomposition is mainly due to the oxidation on semiconductor sulfide minerals (Figure 10); thiosulfate ions are oxidized on the anodic sites of the mineral surface and electrons are transferred through the semiconductor minerals to a cathodic site where the reduction of oxidizing agents such as oxygen and $Cu(NH_3)_4^{2+}$ occurs. In a study by Chen et al. (2008), it was established that pyrite, pyrrhotite, and arsenopyrite can result in thiosulfate decomposition. Based on the mineralogical analysis conducted in this study, HAsBS contains arsenopyrite, pyrite, and pyrrhotite (Figure 1). From this, it can be assumed that the rapid decomposition of thiosulfate with HAsBS is mainly due to the sulfide minerals in the ore. This allowed us to hypothesize that thiosulfate decomposition in the presence of HAsBS may be the dominant mechanism for the suppression of gold extraction. Based on this assumption, a method to enhance the extraction of gold associated with HAsBS is proposed in the next section.

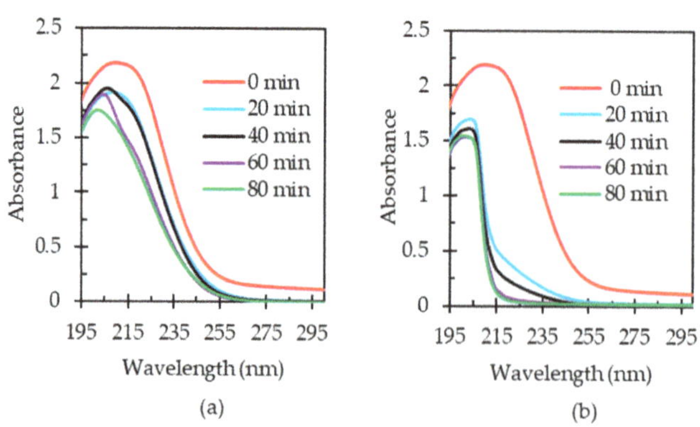

Figure 8. UV–Vis spectra of thiosulfate (**a**) in the absence of HAsBS ore, (**b**) in the presence of HAsBS ore.

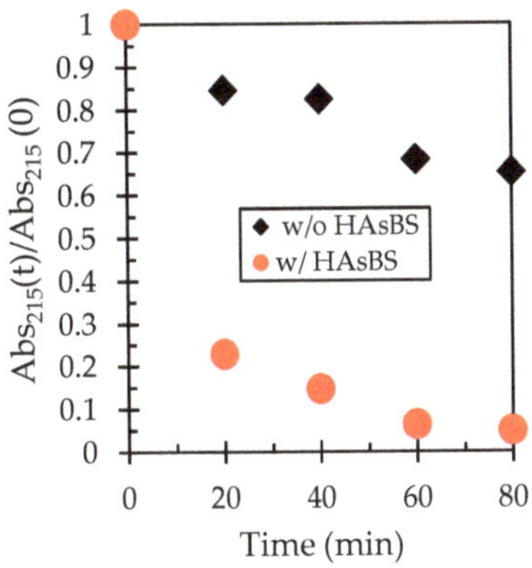

Figure 9. UV−Vis spectra thiosulfate decay of the (Abs_{215}) absorbance for thiosulfate with and without HAsBS.

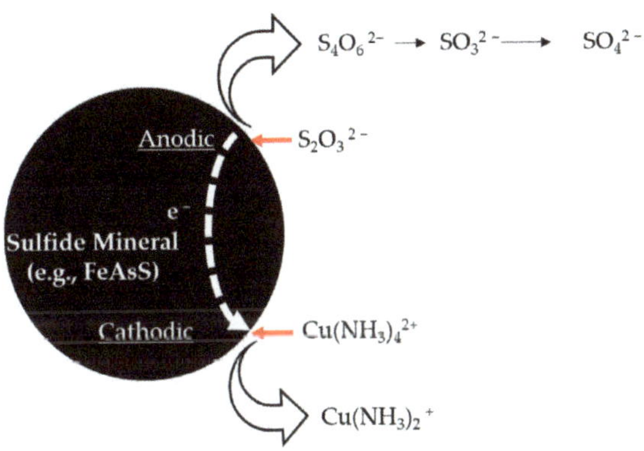

Figure 10. Schematic diagram of the decomposition of thiosulfate on the surface of sulfide minerals [17].

3.5. Pre-Oxidation of HAsBS and Its Effect on Gold Leaching

As discussed in the previous section, the major problem causing the low gold extraction is the decomposition of thiosulfate due to the presence of sulfide minerals in HAsBS. To solve this problem, the present study proposes a two-step process for gold extraction from high arsenopyrite-bearing gold ores: (1) the pre-oxidation of HAsBS using an ammonia solution containing cupric ions, and (2) ammonium thiosulfate leaching of the pretreated ore to extract gold. From Equation (6) and Equation (7), when arsenopyrite and pyrite are oxidized by the cupric ammine complex, the surface of HAsBS will be covered by a "passivation layer" comprised of oxidation products such as ferric arsenate and ferric oxyhydroxides. This passivating material minimizes the oxidative decomposition of thiosulfate ions on the surface of HAsBS, thus enhancing the gold extraction. In this section, demonstration experiments using a model ore composed of gold powder and ground HAsBS were conducted to evaluate the effect of the pre-oxidation on gold extraction.

3.5.1. Pre-Oxidation of HAsBS Using Ammonia Solution Containing Cupric Ions

For the pre-oxidation, a solution containing 0.5 M NH_3, 0.5 M NH_4Cl, and 100 mM $CuCl_2$ was used in which the cupric ammine complex was the oxidant for HAsBS. From Equations (6) and (7), it is expected that iron and arsenic species are precipitated during the oxidation of HAsBS. As shown in Figure 11, after 24 h of pretreatment in an ammonium solution containing cupric ions, about 25% of the sulfur dissolved, but almost no iron and arsenic species were detected in the solution as expected. Because the solution used in the pre-oxidation did not contain thiosulfate, gold extraction in the pre-oxidation step was small (about 1%).

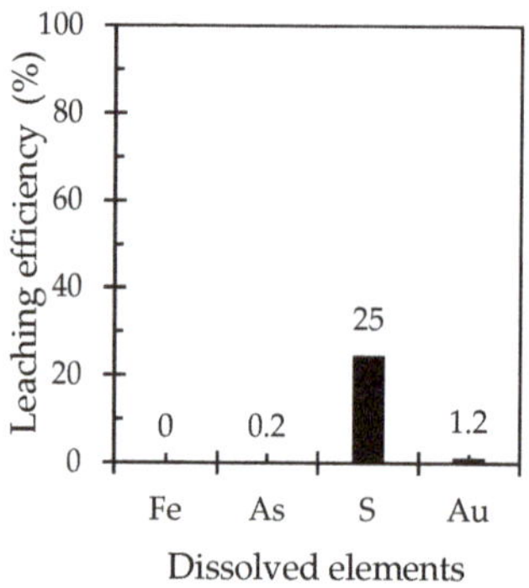

Figure 11. Effects of cupric ammine oxidation on the leaching of HAsBS ore.

The SEM–EDX analysis to evaluate the insulating layer formation on HAsBS after pretreatment for 24 h is shown in Figure 12. The morphology of the HAsBS before and after pre-oxidation shows that the HAsBS surface changed from smooth to "pitted", and it was also confirmed that precipitates were attached to the HAsBS surface after pre-oxidation. From EDX mapping for the residue, Fe, As, S, and O were evenly distributed on the mineral grain. The intensity of oxygen in the EDX spectrum (Figure 13) was higher in the residues as compared to the HAsBS particles before leaching. The stronger oxygen peak for the residue may be attributed to the formation of $FeAsO_4$ adsorbed on the ferric oxyhydroxides precipitated from the oxidation of HAsBS (Equation (8)). Weak copper signals were also detected on the HAsBS leach residues, which might indicate the formation of Cu(I)-sulfide precipitates on the mineral surface as shown in Figure 12(b-5).

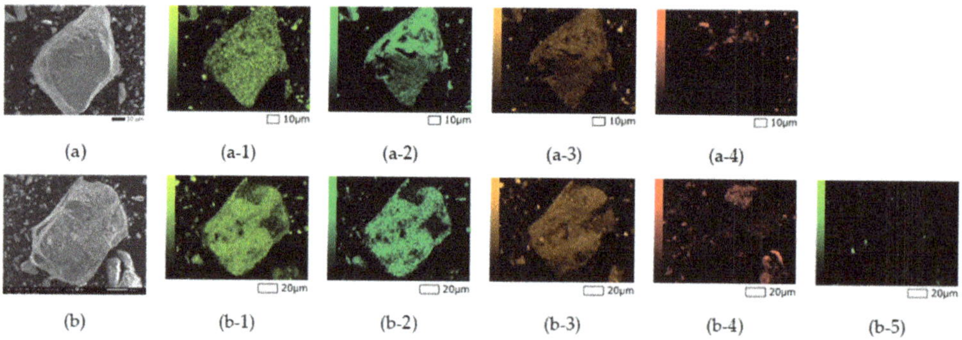

Figure 12. SEM–EDX photomicrographs of (**a**) HAsBS before treatment and the subsequent elemental maps, (**a-1**) Fe, (**a-2**) As, (**a-3**) S, (**a-4**) O; (**b**) pre-oxidized HAsBS; (**b-1**) Fe, (**b-2**) As, (**b-3**) S, (**b-4**) O and (**b-5**) Cu.

Figure 14 shows the XPS results for the residue from pre-oxidation. In the Fe $2p_{3/2}$ spectrum, Figure 14a shows two peaks at 708.2 eV and 711.4 eV, and both were assigned to Fe(III)−O [54,55]. In the deconvoluted results of the As $3d_{5/2}$ spectrum, (Figure 14b), two peaks were detected at 45.6 eV and 47.3 eV, corresponding to As(III)−O and As(V)−O, respectively. In the O 1s spectrum, Figure 14d, on the treated HAsBS, a peak at 532.3 eV, which corresponds to the hydroxyl oxygen (OH^-), was detected. These results are consistent with the hypothesis that ferric arsenate ($FeAsO_4$) and ferric oxyhydroxide (FeOOH) are present as oxidation products of arsenopyrite and pyrite.

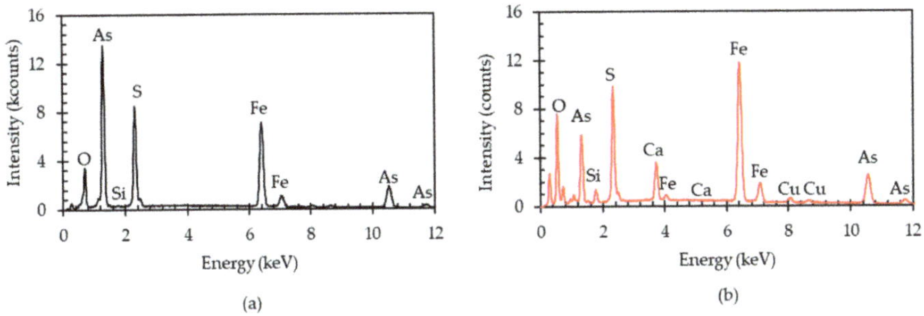

Figure 13. EDX-energy elemental spectra of (**a**) untreated HAsBS, and (**b**) pre-oxidized HAsBS.

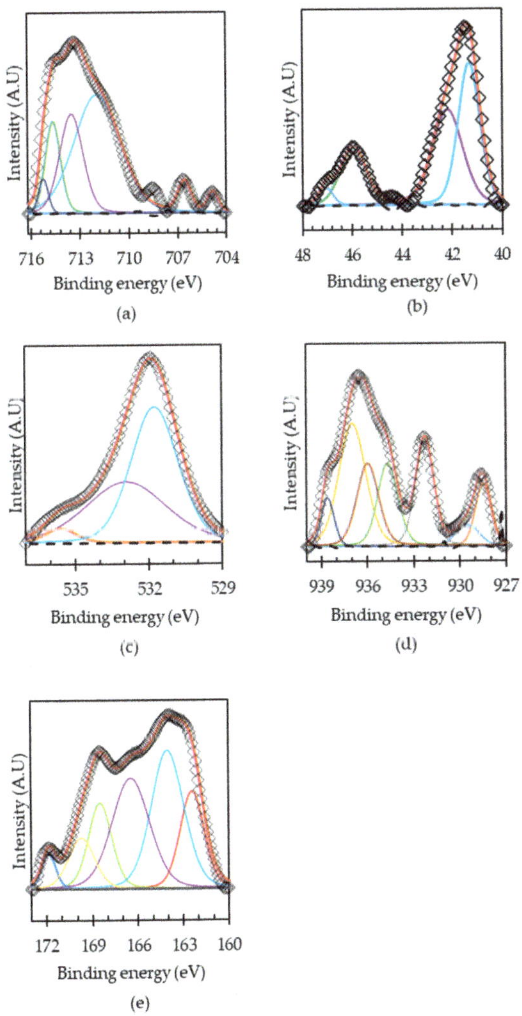

Figure 14. XPS results: (**a**) Fe 2p, (**b**) As 3d, (**c**) O 1s, (**d**) Cu 2p, (**e**) S 2p spectra. The red lines denote the fitting line for the XPS spectra, whilst the green, brown, yellow, purple, and blue lines represent the deconvoluted spectra for the respective elements.

In the deconvoluted S 2p spectrum, Figure 14e, strong peaks at 162.5 eV and 164.1 eV were dominant and assigned to disulfide (S_2^{2-}) attached to Cu(I) species [38], S^{2-} attached to Cu(II) species [56], and elemental sulfur (S) [57], respectively. At higher energy levels, satellite peaks for sulfur were observed at 166.5 eV, 168.6 eV, and 169.8 eV, which are due to sulfate species.

The XPS analysis also confirmed that strong Cu signals appeared after pre-oxidation. In the Cu 2p3/2 spectrum for the residue after pre-oxidation (Figure 14c), a dominant peak was observed at 932.3 eV and this was assigned to Cu(I), which could have originated from Cu_2S or $Cu(S_2O_3)_3^{5-}$ [58]. This may suggest that during the pre-oxidation, copper ions in the solution phase reacted with sulfide minerals in HAsBS ore and formed copper sulfides on the surface. A peak at 934.7 eV corresponded to Cu(II)-oxide species. In addition, the Cu 2p3/2 spectrum had satellite peaks at higher energy levels; that is, 936.0 eV, 937.0 eV, and 938.6 eV, which were assigned to Cu(II)-sulfate species.

The results of surface analyses (SEM–EDX and XPS) confirm that the HAsBS surface is covered by passivating materials such as $FeAsO_4$ and FeOOH after the pre-oxidation step.

3.5.2. Effect of Pre-Oxidation of HAsBS on Thiosulfate Decomposition

The effects of pre-oxidation were assessed in the previous section (Section 3.5.1), and it was found that a passivating layer, which comprised of oxidation products, formed on the surface of HAsBS. In this section, the effect of the pretreatment (pre-oxidation) of HAsBS on thiosulfate decomposition was evaluated using UV–Vis analysis (Figure 15). From the UV–Vis spectra of the thiosulfate solution, which had reacted with untreated HAsBS, the adsorption at 215 nm (Abs_{215}) decreased drastically with time, while the rate of decrease became slower when the thiosulfate solution was reacted with pretreated HAsBS. The decay of absorbance with pretreated HAsBS was almost the same as that without HAsBS. These results confirmed that pre-oxidation of HAsBS, forming a passivating layer on its surface, was effective as a way to minimize the decomposition of thiosulfate.

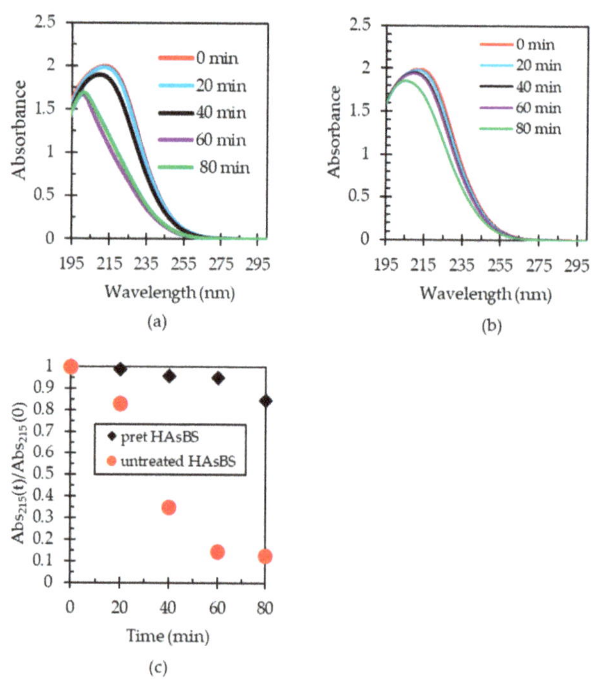

Figure 15. UV–Vis spectra for thiosulfate in the presence of (**a**) untreated HAsBS, (**b**) with pre-oxidized (pret) HAsBS, and (**c**) decay of Abs_{215} value with time.

3.5.3. Effect of Pre-Oxidation of HAsBS on Gold Extraction

Figure 16 shows the effects of the pre-oxidation of HAsBS on the subsequent gold leaching using ammonium thiosulfate solutions, and the results show that gold extraction was significantly improved after pre-oxidation for 24 h; that is, gold extraction was 10% without pre-oxidation, while it increased to 79% when pre-oxidation was applied. The ICP-AES results indicate that passivation of the HAsBS surface was enough to suppress thiosulfate decomposition in the subsequent gold leaching step. When the surface of the HAsBS is oxidized, the oxidation products cover the surface of the sulfides and prevent the transfer of electrons from the thiosulfate ions to the cupric ammine complex as shown in Figure 10, thus minimizing thiosulfate decomposition.

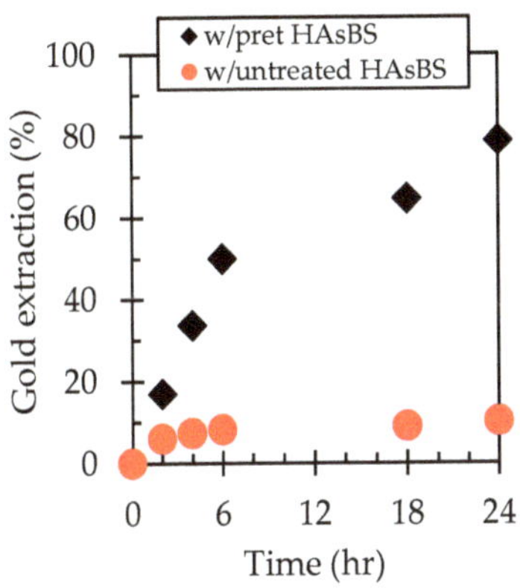

Figure 16. The effect of pre-oxidation of HAsBS on gold extraction.

4. Conclusions

The effects of HAsBS on gold extraction in ammonium thiosulfate were investigated by leaching experiments using a model ore composed of gold powder and ground HAsBS. The results showed that during ammonium thiosulfate leaching of the gold/HAsBS mixture, HAsBS having semiconductor properties facilitated the oxidation of the thiosulfate by the cupric ammine complex, resulting in the decomposition of thiosulfate, which is needed for gold extraction. As a result, gold extraction was suppressed in the presence of HAsBS. The thiosulfate decomposition can be minimized by applying a pre-oxidation treatment of HAsBS in cupric ammonium solutions, which results in passivation of the mineral surface with oxidation products. When the pre-oxidation was applied before ammonium thiosulfate leaching, gold extraction was improved from 10% to 79%. It is further noted that this approach minimizes reagent consumption as the copper (II) ions used in the pretreatment stage can be reused in the subsequent thiosulfate leaching

Author Contributions: Conceptualization, T.J.M., S.J., S.H., I.P., Y.E., M.I. and N.H.; methodology, T.J.M., I.P. and N.H.; investigation, T.J.M. and S.H.; writing—original draft preparation, T.J.M.; writing—review and editing, T.J.M., S.J., I.P., M.I. and N.H.; funding acquisition, S.J. and N.H. All authors have read and agreed to the published version of the manuscript.

Funding: This study was financially supported by the Japan Society for Promotion of Science (JSPS) Grant-in-Aid for Scientific Research (C) (JP23K04642).

Data Availability Statement: Data available on request due to restrictions, as the research is ongoing.

Conflicts of Interest: The authors declare no conflict of interest.

References

1. Ortego, A.; Calvo, G.; Valero, A.; Iglesias-Émbil, M.; Valero, A.; Villacampa, M. Assessment of strategic raw materials in the automobile sector. *Resour. Conserv. Recycl.* **2020**, *161*, 104968. [CrossRef]
2. Moore, J.; James, C. Recent progress and applications of gold nanotechnology in medical biophysics using artificial intelligence and mathematical modeling. *Nano Express* **2021**, *2*, 022001. [CrossRef]
3. WorldGoldCouncil. World Gold Council, GOLDHUB. 1 November 2022. Available online: https://www.gold.org/goldhub/data/gold-supply-and-demand-statistics (accessed on 23 January 2023).
4. Schodde, R. Recent trends in gold discovery. In Proceedings of the 2011 NewGenGold Conference, Perth, Australia, 22–23 November 2011.
5. MetalsFocus. Gold Focus 2020. MetalsFocus. 25 June 2020. Available online: https://www.metalsfocus.com/wp-content/uploads/2020/11/GOLD-FOCUS-2020.pdf (accessed on 14 February 2023).
6. Mudd, G.; Giurco, D.; Mohr, S.; Mason, L. Gold Resources and Production: Australia in a Global Context. The Department of Civil Engineering (Monash University) and the Institute for Sustainable Futures. University of Technology Sydney, Sydney, Australia, 2012. 1 November 2012. Available online: http://cfsites1.uts.edu.au/find/isf/publications/muddetal2012goldresourcesproduction (accessed on 4 December 2022).
7. Tabelin, C.; Park, I.; Jeon, S.; Phengsaart, T.; Villacorte-Tabelin, M.; Alonzo, D.; Yoo, K.; Ito, M.; Hiroyoshi, N. Copper and critical metals production from porphyry ores and E-wastes: A review of resource availability, processing/recycling challenges, socio-environmental aspects, and sustainability issues. *Resour. Conserv. Recycl.* **2021**, *170*, 105610.
8. Vikentyev, I.; Vikent, O.; Tyukova, E.; Nikolsky, M.; Ivanova, J.; Sidorova, N.; Tonkacheev, D.; Abramova, V.; Blokov, V.; Spirina, A. Noble Metal Speciations in Hydrothermal Sulphides. *Minerals* **2021**, *11*, 488. [CrossRef]
9. Donoghue, S.; Young, M. *Environmental Assessment Report*; Dargues Reef Gold Mine Modification 3 (10_0054 MOD 3); NSW Government, Department of Planning & Environment: New South Wales, Australia, 2016.
10. Xiao, L.; Wang, Y.; Qian, P. Advances in non-cyanide process for gold smelting. *Gold Sci. Technol.* **2019**, *27*, 292–301.
11. Adams, C.R.; Porter, C.P.; Robshaw, T.J.; Bezzina, J.P.; Shields, V.R.; Hides, A.; Bruca, R.; Ogden, M.D. An alternative to cyanide leaching of waste activated carbon ash for gold and silver recovery via synergistic dual-lixiviant treatment. *J. Ind. Eng. Chem.* **2020**, *92*, 120–130.
12. Zhang, Y.; Cui, M.; Wang, J.; Liu, X.; Lyu, X. A review of gold extraction using alternatives to cyanide: Focus on current status and future prospects of the novel eco-friendly synthetic gold lixiviants. *Miner. Eng.* **2022**, *176*, 107336. [CrossRef]
13. Brooy, S.; Linge, H.; Walker, G. Review of gold extraction from ores. *Miner. Eng.* **1994**, *70*, 1231–1241. [CrossRef]
14. Celep, O.; Altinkaya, E.; Yazici, Y.; Deveci, H. Thiosulfate leaching of silver from arsenical refractory ore. *Miner. Eng.* **2018**, *122*, 63–70. [CrossRef]
15. Jeon, S.; Ito, M.; Tabelin, C.B.; Pongsmrankul, R.; Kitajima, N.; Park, I.; Hiroyoshi, N. Gold recovery from shredder light fraction of E-waste recycling plant by flotation-ammonium thiosulfate leaching. *Waste Manag.* **2018**, *77*, 195–202. [CrossRef]
16. Xie, F.; Chen, J.; Wang, J. Review of gold leaching in thiosulfate-based solutions. *Trans. Nonferrous Met. Soc. China* **2021**, *31*, 3506–3529. [CrossRef]
17. Xu, B.; Wenhao, K.; Qian, L.; Yongbin, Y.; Tao, J.; Liu, X. A Review of Thiosulfate Leaching of Gold: Focus on Thiosulfate Consumption and Gold Recovery from refractory ores. *Minerals* **2017**, *7*, 222.
18. Jeon, S.; Tabelin, C.B.; Takahashi, H.; Ilhwan, P.; Ito, M.; Hiroyoshi, N. Enhanced cementation of gold via galvanic interactions using activated arbon and zero-valent aluminum: A novel approach to recover gold ions from ammonium thiosulfate medium. *Hydrometallurgy* **2020**, *191*, 105165. [CrossRef]
19. Feng, D.; Van Deventer, J. Leaching behaviour of sulphides in ammoniacal thiosulphate systems. *Hydrometallurgy* **2002**, *63*, 5–13. [CrossRef]
20. Feng, D.; Van Deventer, J. The role of heavy metal ions in gold dissolution in the ammoniacal thiosulfate system. *Hydrometallurgy* **2002**, *64*, 231–246. [CrossRef]
21. Yang, Y.; Zhang, X.; Xu, B.; Li, Q.; Jiang, T.; Wang, Y. Effect of arsenopyrite on thiosulfate leaching of gold. *Trans. Nonferrous Met. Soc. China* **2015**, *25*, 3454–3460. [CrossRef]
22. Park, I.; Higuchi, K.; Tabelin, C.; Jeon, S.; Ito, M.; Hiroyoshi, N. Suppression of arsenopyrite oxidation by microencapsulation using ferric-catecholate complexes and phosphate. *Chemosphere* **2021**, *269*, 129413. [CrossRef]
23. Nie, Y.; Chen, J.; Wang, Q.; Zhang, C.; Shi, C.; Zhao, J. Use of dry grinding process to increase the leaching of gold from a roasted concentrate containing hematite in the thiosulfate system. *Hydrometallurgy* **2021**, *201*, 105582. [CrossRef]
24. Bowell, R.; Alpers, C.; Jamieson, H.; Nordstrom, D.; Majzlan, J. The environmental geochemistry of arsenic- an overview. *Rev. Mineral. Geochem.* **2014**, *79*, 1–16. [CrossRef]
25. Yang, D.; Shi, M.; Zhang, J.; Sasaki, A.; Endo, M. Reductive roasting of arsenic-contaminated red mud for Fe resources recovery driven by johnbaumite-based arsenic thermostabilization strategy. *J. Harzadous Mater.* **2023**, *452*, 131255. [CrossRef]
26. Schlesinger, W.H.; Klein, E.M.; Vengosh, A. The Global Biogeochemical Cycle of Arsenic. *Glob. Biogeochem. Cycles* **2022**, *36*, e2022GB007515. [CrossRef]
27. Duan, L.; Li, X.; Jiang, Y.; Lei, M.; Dong, Z. Arsenic transformation behaviour during thermal decomposition of P. vittata, an arsenic hyperaccumulator. *J. Anal. Appl. Pyrolysis* **2017**, *124*, 584–591. [CrossRef]

28. Fleming, C.; McMullen, J.; Thomas, K.; Wells, J. Recent advances in the development of an alternative to the cyanidation process: Thiosulfate leaching and resin in pulp. *Min. Met. Explor.* **2001**, *20*, 1–9. [CrossRef]
29. Grosse, A.; Dicinoski, G.; Shaw, M.; Haddad, P. Leaching and recovery of gold using ammoniacal thiosulfate leach liquors. *Hydrometallurgy* **2003**, *69*, 1–21. [CrossRef]
30. Zhang, X.; Senanayake, G. A review of ammoniacal thiosulfate leaching of gold: An update useful for further research in Non-cyanide gold lixiviants. *Miner. Process. Extr. Metall. Rev.* **2016**, *37*, 385–411. [CrossRef]
31. Ter-Arakelyan, K. On technological expediency of sodium thiosulfate usage for gold extraction from raw material. *Izv. V. U. Z. Tsvetn. Metall.* **1984**, *12*, 72–76.
32. Alguacil, F. *The Chemistry of Gold Extraction*, 2nd ed.; John O Marsden and Clain House SME: Littleton, CO, USA, 2006.
33. Salinas-Rodríguez, E.; Hernández-Ávila, J.; Cerecedo-Sáenz, E.; Arenas-Flores, A.; Veloz-Rodríguez, M.; Toro, N.; Gutiérrez-Amador, M.; Acevedo-Sandoval, O.A. Leaching of Copper Contained in Waste Printed Circuit Boards, Using the Thiosulfate—Oxygen System: A Kinetic Approach. *Materials* **2022**, *15*, 2354. [CrossRef]
34. Aylmore, M.; Muir, D. Thiosulfate leaching of gold-A review. *Miner. Eng.* **2001**, *14*, 135–174. [CrossRef]
35. Robert, T.; Bartel, M.; Offergeld, G. Characterization of oxygen species adsorbed on copper and nickel oxides by X-ray photoelectron spectroscop. *Surf. Sci.* **1972**, *33*, 123–130. [CrossRef]
36. Nakai, I.; Sugitani, Y.; Nagashima, K.; Niwa, Y. X-ray photoelectron spectroscopic study of copper minerals. *J. Inorg. Nucl. Chem.* **1978**, *40*, 789–791. [CrossRef]
37. Mielczarski, J.; Minni, E. The adsorption of diethyldithiophosphate on cuprous sulphide. *Surf. Interface Anal.* **1984**, *6*, 221. [CrossRef]
38. Wagner, C.; Riggs, W.; Davis, L.; Moulder, J.; Muilenberg, G. *Handbook of X-ray Photoelectron Spectroscopy*; Perkin-Elmer Corporation, Physical Electronics Division: Eden Prairie, MI, USA, 1979.
39. Hitchcook, A.; Brion, C. K-shell excitation spectra of CO, N_2 and O_2. *J. Electron Spectrosc. Relat. Phenom.* **1980**, *18*, 1–21. [CrossRef]
40. Nefedov, V.; Gati, D.; Dzhurinskii, B.; Sergushin, N.; Salyn, Y. A qualitative and quantitative study of the oxides of aluminum and silicon using AES and XPS. *Russ. J. Inorg. Chem.* **1975**, *35*, 1279.
41. Can, F.; Demirci, O.; Dumoulin, F.; Erhan, E.; Arslan, L.; Ergenekon, P. Iron porphyrin-modified PVDF membrane as a biomimetic material and its effectiveness on nitric oxide binding. *Appl. Surf. Sci.* **2017**, *420*, 625–630. [CrossRef]
42. Rajumon, M.; Prabhakaran, K.; Rao, C. Adsorption of oxygen on (100), (110) and (111) surfaces of Ag, Cu and Ni: An electron spectroscopic study. *Surf. Sci.* **1990**, *233*, 233–242. [CrossRef]
43. Park, I.; Tabelin, C.; Seno, K.; Jeon, S.; Inano, H.; Ito, M.; Hiroyoshi, N. Carrier-microencapsulation of arsenopyrite using Al-catecholate complex:nature of oxidation products, effects on anodic and cathodic reactions, andcoating stability under simulated weathering condition. *Heliyon* **2020**, *6*, 03189. [CrossRef]
44. Chen, M.; Zhang, Z.; Hu, X.; Tian, J.; Wang, J.; Wan, R.; Zhou, X.; Shen, P.; Liu, D. Oxidation mechanism of the arsenopyrite surface by oxygen with and without water: Experimental and theoretical analysis. *Appl. Surf. Sci.* **2022**, *573*, 151574.
45. Zhu, T.; Lu, X.; Liu, H.; Li, J.; Zhu, L.; Lu, J.; Wang, R. Quantitative X-ray photoelectron spectroscopy-based depth profiling of bioleached arsenopyrite surface by Acidithiobacillus ferrooxidans. *Geochim. Et Cosmochim. Acta* **2014**, *127*, 120–139. [CrossRef]
46. Feng, X.; Liu, Q.; Wang, S.; Cen, L.; Li, H. Arsenopyrite weathering in acid rain: Arsenic transfer and environmental implications. *J. Hazard. Mater.* **2021**, *420*, 126612. [CrossRef]
47. Hollinger, G.; Kabban, I.S.; Gendry, M. Oxides on GaAs and InAs surfaces: An x-ray-photoelectron-spectroscopy study of reference compounds and thin oxide layers. *Phys. Rev.* **1994**, *49*, 11159. [CrossRef]
48. Bilal, M.; Ito, M.; Akashino, R.; Bu, X.; Hassan, F.; Park, I.; Jeon, S.; Aikawa, K.; Hiroyoshi, N. Heterogenous carrier flotation technique for recovering finely ground chalcopyrite particles using coarse pyrite particles as a carrier. *Miner. Eng.* **2022**, *180*, 107518. [CrossRef]
49. Wang, X.; Wang, J.; Zhang, X.; Tian, Q.; Liu, M.; Cai, N.; Xue, Y.; Chen, W.; Li, W.; Yu, F. Nitrogen-Doped Cu_2S/MoS_2 Heterojunction Nanorod Arrays on Copper Foam for Efficient Hydrogen Evolution Reaction. *ChemCatChem* **2018**, *11*, 1143–1373. [CrossRef]
50. Wu, C.; Yin, M.; O'Brien, S.; Koberstein, J. Quantitative Analysis of Copper Oxide Nanoparticle Composition and Structure by X-ray Photoelectron Spectroscopy. *Chem. Mater.* **2006**, *18*, 6054–6058. [CrossRef]
51. Biesinger, M.; Lau, L.; Gerson, A.; Smart, R. Resolving surface chemical states in XPS analysis of first row transition metals, oxides and hydroxides: Sc, Ti, V, Cu and Zn. *Appl. Surf. Sci.* **2010**, *257*, 887–898.
52. Khan, S. UV-ATR Spectroscopy Study of the Speciation in Aqueous Polysulfide Electrolyte Solutions. *Int. J. Electrochem. Sci.* **2012**, *7*, 561–568. [CrossRef]
53. Pappoe, M.; Bottaro, C. Systematic optimization of pyromellitic acid background electrolyte for capillary electrophoresis with indirect UV-Vis detection and online pre-concentration analysis of thiosalt anions in treated mine tailing. *Anal. Methods* **2014**, *6*, 9305–9312.
54. Liu, X.; Li, Q.; Zhang, Y.; Jiang, T.; Yang, Y.; Xu, B.; He, Y. Electrochemical behaviour of the dissolution and passivation of arsenopyrite in 9K culture medium. *Appl. Surf. Sci.* **2020**, *508*, 145269. [CrossRef]
55. de Jong, A.B.H.; Van Ijzendoorn, L.; Soudant, V.; de Beer, V. Preparation, Structure and Surface Chemical Properties of Hydrotreating Model Catalysts: A Surface Science Approach. *J. Phys. Chem.* **1993**, *97*, 6477.

56. Yu, X.-R.; Liu, F.; Wang, Z.-Y.; Chen, Y. Auger parameters for sulfur-containing compounds using a mixed aluminum-silver excitation source. *J. Electron Spectrosc. Relat. Phenom.* **1990**, *50*, 159–166. [CrossRef]
57. Contini, G.; Laajalehto, K.; Suoninen, E.; Marabini, A. 5-Methyl-2-mercaptobenzoxazole adsorbed onto chalcocite (Cu_2S): An XPS and X-AES study. *J. Colloid Interface Sci.* **1995**, *171*, 234–239.
58. Feng, D.; Van Deventer, J. The effect of sulfur species on thiosulfate leaching of gold. *Miner. Eng.* **2007**, *20*, 273–281. [CrossRef]

metals

MDPI

Article

Selective Cementation of Gold Using an Iron Oxide and Zero-Valent Aluminum Galvanic System from Gold–Copper Ammoniacal Thiosulfate Solutions

Joshua Zoleta [1,2,*], Sanghee Jeon [3], Akuru Kuze [1], Nako Okada [1], Ilhwan Park [4], Mayumi Ito [4], Yogarajah Elakneswaran [4] and Naoki Hiroyoshi [4,*]

1 Laboratory of Chemical Resources, Division of Sustainable Resources Engineering, Graduate School of Engineering, Hokkaido University, Sapporo 060-8628, Japan; n_okada0116@eis.hokudai.ac.jp (N.O.)
2 Department of Materials and Resources Engineering, College of Engineering and Technology, Mindanao State University-Iligan Institute of Technology, Iligan City 9200, Philippines
3 Department of Earth Resource Engineering and Environmental Science, Faculty of International Resources Science, Akita University, Akita 010-0865, Japan
4 Division of Sustainable Resources Engineering, Faculty of Engineering, Hokkaido University, Sapporo 060-8628, Japan
* Correspondence: joshua.zoleta@g.msuiit.edu.ph (J.Z.); hiroyosi@eng.hokudai.ac.jp (N.H.); Tel.: +81-639-1715-80601 (J.Z.)

Abstract: Ammonium thiosulfate leaching is a promising alternative to the conventional cyanide method for extracting gold from ores. However, strategies for recovering gold from the leachate are less commercially used due to its low affinity to gold. The present study investigated the recovery of gold from the leachate using iron oxides (hematite, Fe_2O_3 or magnetite, Fe_3O_4). Cementation experiments were conducted by mixing 0.15 g of aluminum powder as an electron donor and 0.15 g of an electron mediator (activated carbon, hematite, or magnetite) in 10 mL of ammonium thiosulfate leachate containing 100 mg/L gold ions and 10 mM cupric ions for 24 h at 25 °C. The results of the solution analysis showed that when activated carbon (AC) was used, the gold was recovered together with copper (recoveries were 99.99% for gold and copper). However, selective gold recovery was observed when iron oxides were used, where the gold and copper recoveries were 89.7% and 21% for hematite and 85.9% and 15.4% for magnetite, respectively. An electrochemical experiment was also conducted to determine the galvanic interaction between the electron donor and electron mediator in a conventional electrochemical setup (hematite/magnetite–Al as the working electrode, Pt as the counter electrode, Ag/AgCl as the reference electrode) in a gold–thiosulfate medium. Cyclic voltammetry showed a gold reduction "shoulder-like" peak at −1.0 V using hematite/Al and magnetite/Al electrodes. Chronoamperometry was conducted and operated at a constant voltage (−1.0 V) determined during cyclic voltammetry and further analyzed using SEM-EDX. The results of the SEM-EDX analysis for the cementation products and electrochemical experiments confirmed that the gold was selectively deposited on the iron oxide surface as an electron mediator.

Keywords: ammoniacal thiosulfate leaching; reductive precipitation; galvanic interactions; cyclic voltammetry; chronoamperometry; iron oxides

1. Introduction

Recently, high-grade gold ores have been reported to be gradually depleting, and the refractory gold ores comprising approx. one-third of the total gold production from open pit/underground mines and in tailing deposits have become the interest of this study. In refractory ores, sulfide-type gold ores, especially gold-pyritic ore, account for a substantial proportion of gold resources (i.e., about 22%). The recovery of gold from gold-pyritic ore has recently come into the spotlight as a valuable study in resources fields [1–4]. Additionally, gold recovery places environmental hazards on storing these sulfide-rich minerals, as these minerals potentially form in highly acidic water, commonly known as acid mine drainage (AMD) [4]. Currently, for processing refractory gold ore, conventional processing is conducted through the roasting process followed by the cyanidation process. This current technology is being practiced in industrial mining plants, such as the roasting–cyanidation leaching plant of Zhongyuan Gold Smelter Co. Ltd. in Henan Province, Sanmenxia, China [5]. In the later leaching process, where cyanide is conventionally employed, the solvent exhibits health and environmental hazards due to its potential toxicity if not appropriately managed [1,2,6]. For these reasons, many studies have been conducted to find alternative solvents to mitigate the risk from using cyanide, such as thiourea [7–9], thiocyanate [10], halogen-based [11–14] and thiosulfate [15–21] solvents. Among the alternatives, ammonium thiosulfate has been highlighted since it has been shown to have a higher selectivity and leaching rate to gold, and most importantly, it is less corrosive and non-toxic to humans [15–21]. Methods to recover gold from the leachate have not been extensively practiced due to a low affinity to carbon during the gold adsorption recovery process [16–18,22–25], which has caused limitations in commercial operations.

For these reasons, the development of an alternative recovery method for gold ions in the ammonium thiosulfate system is needed. A recent study reported that gold recovery could be efficiently achieved by cementing the gold ions via galvanic interactions between activated carbon (electron mediator) and zero-valent aluminum (ZVAl) (electron donor) in a gold–copper ammoniacal thiosulfate system [15]. The recovery results using ZVAl or AC in a single system were negligible. However, when mixed as a binary system, over 99% of the Au ions could be recovered through the following suggested mechanisms, where ZVAl acted as an electron donor (i.e., anode) and activated carbon (AC) acted as an electron mediator (i.e., cathode), creating the galvanic cell and finally leading to a reductive deposition of the gold–thiosulfate complex onto the activated carbon attached to ZVAl [16–18]. To further explore the potential of this technique, further researched was conducted to understand the behavior of gold ions in an ammonium thiosulfate system in the presence of various metallic combinations of electron donors, zero-valent aluminum, and electron mediators, such as Fe and Cu. The results showed that approx. 20% and 40% of gold ions were recovered, respectively [15]. Although the recovery of gold ions was not discussed, Choi et al. also further conducted a study to remove cadmium (Cd^{2+}) and zinc (Zn^{2+}) ions from the sulfate-based solvents using aluminum and iron oxide materials (e.g., Fe_3O_4). The results showed that approx. 83% of Cd^{2+} and 92% of Zn^{2+} were removed from the acid solutions. In contrast, the recovery efficiency was negligible when employing the ZVAl sole system (i.e., 0% recovery of Cd^{2+} and Zn^{2+}) due to the thin insulating oxyhydroxide film on the surface of the ZVAl. Although the recovery efficiencies were lower than the Al/AC system, the research highlighted that metal ions could be recovered via galvanic interactions using ZVAl and other conductive materials [26].

As previously mentioned, when dealing with pyritic refractory ores, the leaching agent cannot contact gold in the ore because the gold grains are encapsulated in sulfide minerals such as pyrite, FeS_2 [27]. Amongst all the pretreatment processes, roasting has been conventionally employed to convert pyrite into various porous iron oxides, such as hematite (Fe_2O_3) and magnetite (Fe_3O_4), so that the gold grains can be exposed to the leach solutions, making them susceptible for extraction [3,4,28–30].

Considering the semiconductive properties of hematite (Fe_2O_3) and magnetite (Fe_3O_4), it is possible for them to be used as electron mediators in the cementation process proposed by the previous researchers [15–18]. In this case, activated carbon can be eliminated when processing refractory gold ores. By treating refractory gold in pyrite (FeS_2), we could mitigate the formation of acid mine drainage (AMD). In the present study, the technical feasibility of two iron oxides (hematite (Fe_2O_3) and magnetite (Fe_3O_4); common products for pyrite roasting) as the electron mediators for the enhanced cementation of gold from ammonia thiosulfate leachate was evaluated using zero-valent aluminum as an electron donor.

2. Materials and Methods

2.1. Materials

All the chemicals used were purchased from Wako Pure Chemical Industries, Ltd., Osaka, Japan. ZVAl (99.99%, CAS No.: 012-19172) was used as an electron donor in the cementation experiments. Activated carbon (99.99%, CAS No.: 031-02135) with approx. 800–1500 m^2/g of a specific surface area, hematite (Fe_2O_3, 99.99%, CAS No.: 096-02821), and magnetite (Fe_3O_4, 99.99%, CAS No.: 093-01035) were used as the electron mediators.

2.2. Solution Preparation

The Au–ammonium thiosulfate solution was prepared by leaching 50 mg of Au powder (99.999%, CAS No.: 937902 Wako Pure Chemical Industries, Ltd., Osaka, Japan) in 500 mL of an ammonium thiosulfate solution containing 1 M $Na_2S_2O_3$ $5H_2O$ (CAS No.: 197-03585), 0.5 M NH3 (CAS No.: 016-03146), 0.25 M $(NH_4)_2SO_4$ (CAS No.: 016-03445), and 10 mM $CuSO_4$ (CAS No.: 034-04445). The materials were mixed in a beaker using a magnetic stirrer with a built-in heater and a part of a thermocouple was submerged in the solution to maintain a temperature of 30 °C for 24 h with a constant shaking frequency of 600 min^{-1}. Electrolyte solutions were prepared by dissolving $Na_2S_2O_3$ $5H_2O$ and metal ions (Cu^{2+} and Au^+) into 0.1 M NH_3/0.05 M $(NH_4)_2SO_4$ buffer solutions.

2.3. Cementation Experiments

The cementation experiments were conducted by mixing 0.15 g of the electron donor (ZVAl) and 0.15 g of an electron mediator (AC, Fe_2O_3, or Fe_3O_4) with 10 mL of a gold–ammonium thiosulfate solution in 50 mL Erlenmeyer flasks using a thermostated water bath shaker (shaking amplitude, 40 mm; frequency, 120 min^{-1}) at 25 °C. Most of the experiments were conducted in the air, while a part was conducted by purging oxygen in the flask and solution using ultra-pure nitrogen (N_2) gas. After a predetermined time, the filtrates and residues were separated via filtration using 0.2 μm syringe-driven membrane filters (LMS et al., Tokyo, Japan).

The residues were thoroughly washed with deionized water, followed by drying the residues in a vacuum oven at 40 °C, and then further analyzed using scanning electron microscopy with energy-dispersive X-ray (SEM-EDS, JSM IT200TM, JEOL Co., Ltd., Tokyo, Japan), operated at an accelerating voltage of 15 kV, 1000× to 1500× magnification and a working distance (WD) of 10 mm. The elemental maps were produced at 2000 cps with a 60-min time constant and a high pixel resolution of 256 × 256 (~10-min scans).

The concentrations of gold ions that remained in the sterile solution were determined using inductively coupled plasma atomic emission spectroscopy (ICP-AES) (ICPE-9820, Shimadzu Corporation, Kyoto, Japan, with a margin of error = ±2%). The Au recovery (Au_R) was then calculated according to the following equation.

$$Au_R = \frac{[Au_i] - [Au_f]}{[Au_i]} \times 100 \quad (1)$$

where $[Au_i]$ and $[Au_f]$ are denoted as the initial and final dissolved Au concentrations, respectively. Note that the initial and final dissolved gold concentrations were measured, and the recovery experiments were performed in three replicate samples.

2.4. Electrochemical Measurements

2.4.1. Preparation of the Aluminum–Hematite (Al/Hem) and Aluminum–Magnetite Electrode (Al/Mag)

The improvised Al electrode was prepared by cutting a cuboid (10 ×10 × 2 mm) from of a large aluminum metal sheet using a diamond cutter, connecting it to copper wires using silver conducting paste, and fixing it inside a plastic mold (25 mm diameter and 10 mm height) using Technovit® non-conductive resin. Next, the surface of the Al electrode was exposed by polishing it with silicon carbide papers of decreasing grain size (#200, #600, #1000, #1500), followed by polishing it with 5 and 1 μm Al_2O_3 pastes on a smooth glass plate. Finally, the polished electrode was ultrasonically cleaned for 5 min to remove residually attached Al_2O_3 particles and washed several times with DI water. The exposed surface of the polished Al working electrode was then oxidized under atmospheric conditions for 2 days to mimic the oxide layer on Al during the cementation process. The oxidized Al working electrode was subjected to SEM-EDX analysis using SEM-EDS, JSM-IT200TM, JEOL Co., Ltd., Tokyo, Japan to elucidate the magnitude of the oxide being formed. The iron oxide powder attached to the Al electrodes (iron oxide/Al electrodes) was then prepared by mixing 0.05 g of iron oxide powder (Fe_3O_4 or Fe_2O_3) with 5 mL of acetone, and the resulting mixture was then carefully attached to the surface of the Al electrode. After the acetone was evaporated, the iron oxide/Al electrodes were used as the working electrode. This procedure was chosen to prevent any "scratching" of the Al oxide layer formed on the electrode and has been successfully used in previous electrochemical studies [31].

Before conducting the electrochemical measurements of the improvised FexOX/Al working electrodes, backscattered electron photomicrographs with elemental mapping and point EDX analysis of the aluminum electrode were taken, as shown in Figure 1. The result showed that traces of O were detected after exposing the surface of the aluminum electrode for 2 days. These traces were dispersedly formed on the surface of the electrode. The presence of this layer mimicked the surface oxide layer of ZVAl during the gold recovery process. Thus, iron oxides were added to the Al electrode surface to understand the electrochemical properties of hematite and magnetite on the Au recovery.

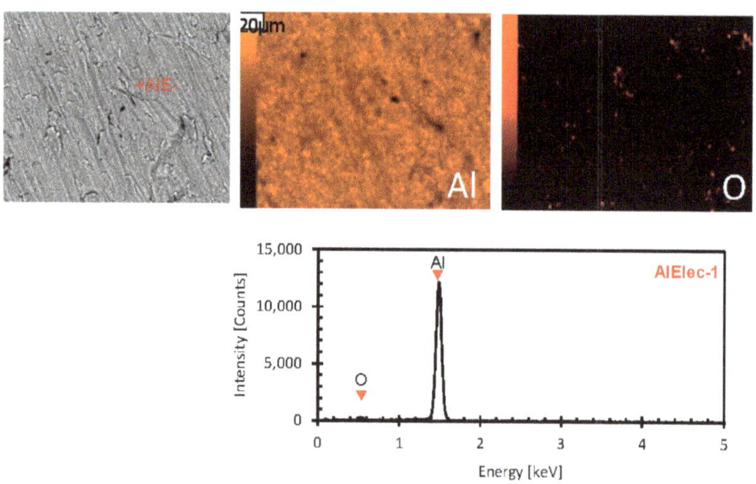

Figure 1. Backscattered electron photomicrograph with elemental mapping and point EDS analysis of the improvised aluminum electrode.

2.4.2. Cyclic Voltammetry Measurements

To elucidate the electrochemical properties of the iron oxide/Al electrodes, cyclic voltammetry (CV) was conducted using a computer-driven potentiostat (SP-300, Biologic, Vaucanson, France) with a conventional three-electrode system composed of an iron oxide/Al electrode as the working electrode, an Ag/AgCl electrode filled with saturated KCl as the reference electrode, and a platinum (Pt) electrode as the counter electrode. The measurements were conducted in an anoxic condition (without dissolved oxygen) by purging the prepared 150 mL electrolyte solution with ultra-pure N_2 gas (99.99%) for 45 min before the electrochemical measurements were taken. Three electrodes were immersed in a glass cell with a water jacket before the N_2 purge and equilibrated at 25 °C for 30 min. In all the cases, the measurements were always taken after the equilibration of the working electrode to its open circuit potential (OCP). Equilibrium, here, means that the measured electrode potential of the working electrode did not change by more than 2 mV for 60 s. After that, the scan began from the OCP and moved towards more negative potentials at a rate of 20 mV/s up to −1.5 V, after which the sweep direction was reversed and moved towards increasingly positive potentials. The scan direction was again reversed after reaching +1.0 V and then moved back to the starting position (i.e., OCP). This entire process constituted one cycle. Each measurement lasted three cycles under unstirred conditions. After each experiment, the electrode was then re-polished using fine-grained Si–carbide papers (#1200 and #1500) and Al_2O_3 pastes (5 µm and 1 µm) to expose a new and unreacted surface for the next cycle, and 5 mg of new iron oxide particles were attached to the surface of the triplicate measurements using this technique.

2.4.3. Chronoamperometry and Electrode Surface Characterization

The electrodes for the surface characterization were prepared by applying the chronoamperometry technique using a computer-driven potentiostat electrochemical measurement (SP-300, Biologic, Vaucanson, France) with a conventional three-electrode system at 25 °C. After the equilibration of the working electrode to the OCP, the electrode was polarized at a fixed electrode potential of −1.0 V for 10 min using Au and Cu electrolytes. This potential was selected since the reduction peak of gold was observed using cyclic voltammetry. The process was agitated at a constant speed of 120 rpm using a magnetic stirrer. After the process, the working electrode was removed from the cell, immersed in deionized water, and cured in a vacuum oven at 40 °C for 24 h. The surface of the working electrode was then investigated using scanning electron microscopy with energy-dispersive X-ray, SEM-EDX [31].

3. Results and Discussion

3.1. Recovery of Gold Ions from Ammonium Thiosulfate Solution

Figure 2 shows the percent recovery of gold ions from a gold–copper ammoniacal thiosulfate solution using different galvanic constituents: (a) ZVAl/and or activated carbon (AC), (b) ZVAl/and or hematite (Hem), and (c) ZVAl/ and or magnetite (Mag) in the air, respectively. Figure 2a shows that the gold recovery was negligible when ZVAl was used without electron mediators, such as AC, due to the passivation of the ZVAl surface with aluminum oxide insulator. This passivation layer (aluminum oxide layer) hindered the transfer of electrons from ZVAl to the gold ions in the solution phase, suppressing the electrochemical deposition of gold and copper [15–18]. Activated carbon had a high specific surface area. It functioned as an adsorbent for the metal ions even without electron donors, such as ZVAl, while the gold and copper recoveries with activated carbon were approx. 24% and 5%, respectively. The low recoveries indicated a limited adsorption of gold and copper complexes from the aqueous phase to activated carbon [16–18]. The copper and gold recoveries were also deficient even when activated carbon and ZVAl were used together. As shown in Figure 2b for hematite and Figure 2c for magnetite, it was also found that the gold and copper recoveries were limited when using iron oxides alone or iron oxide and ZVAl together.

The low recoveries of gold at 24 h in the air (Figure 2) may have been due to the dissolution of cemented gold caused by oxidation with O_2 [16–18]. Figure 3 shows the recoveries of gold using ZVAl and electron mediators as a function of time. It was observed that the gold recoveries increased with time to reach a maximum of 58% at the first 1 h for AC/Al, 75% at 3 h for Fe_2O_3/Al, and 84% at 6 h for Fe_3O_4/Al. All the gold recoveries then decreased with time. The increase in the gold recoveries in the initial period suggested that gold was deposited in the presence of both ZVAl and the electron mediators in the short term due to the following equation.

$$2Au(S_2O_3)_2^{3-} + 6Al^0 + 6H^+ = 2Au^0 + 8S^0 + 3Al_2O_3 + 3H_2O \quad (2)$$

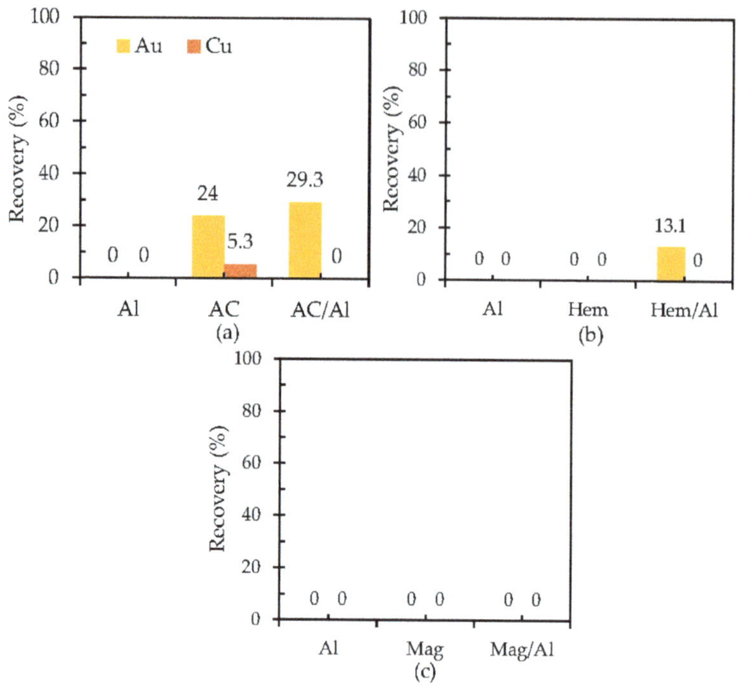

Figure 2. Percent recovery of gold and copper from the different galvanic constituents in an air atmosphere using (**a**) activated carbon (AC) and/or ZVAl, (**b**) hematite (Hem) and/or ZVAl, and (**c**) magnetite (Mag) and/or ZVAl.

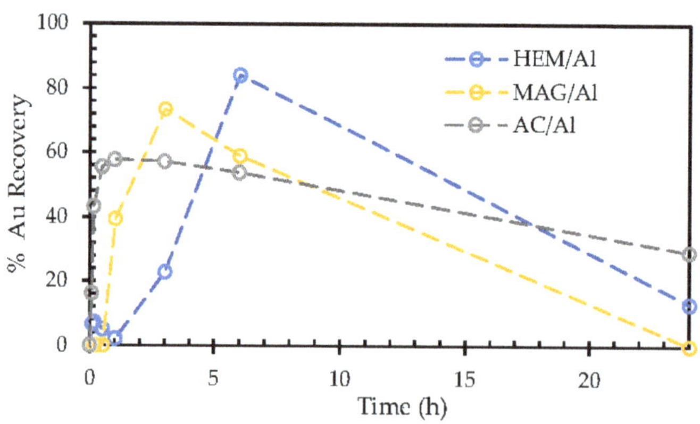

Figure 3. Percent recovery of gold with respect to time using ZVAl and different electron mediators.

However, the decrease in the gold recoveries after the maxima, as observed in Figure 3, may have been due to the oxidation of the deposited gold with dissolved oxygen, as shown in Equation (3).

$$4Au^0 + 8(S_2O_3)^{2-} + O_2 + 2H_2O = 4Au(S_2O_3)_2{}^{3-} + 4OH^- \quad (3)$$

Since the dissolution of gold was significantly affected in the presence of oxygen (O_2), the subsequent cementation experiments were conducted without O_2 by purging the gold thiosulfate solution with N_2 gas for 45 min, as presented in Figure 4a–c, and using ZVAl as an electron donor and various electron mediators. When ZVAI was used without electron mediators, the gold and copper recoveries were slightly improved. The recoveries were 35% for Au and 29% for Cu. These results may have been due to the limited passivation of the Al_2O_3 layer on the ZVAl surface, allowing for a limited electron transfer from the electron donor (ZVAl) to the metal ions. Insignificant gold and copper recoveries were observed when only electron mediators were used (without ZVAl). The recoveries for Au and Cu were less than 40%, regardless of the electron mediator. This result could indicate that the ability of AC, Fe_2O_3, and Fe_3O_4 as adsorbents to the metal ions is limited.

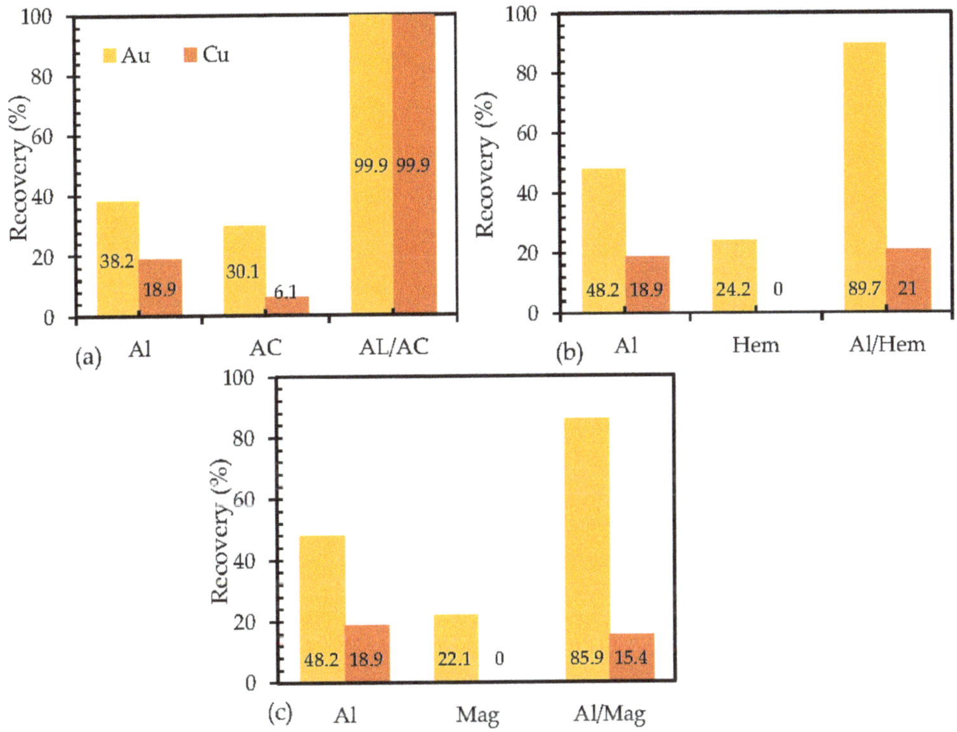

Figure 4. Percent recovery of gold and copper from different galvanic constituents in a nitrogen atmosphere using (**a**) activated carbon (AC) and/or ZVAl, (**b**) hematite (Hem) and/or ZVAl, and (**c**) magnetite (Mag) and/or ZVAl.

However, when both ZVAl as an electron donor and the electron mediators were used together (binary system), high gold recoveries were observed. The recoveries were approx. 99% using AC/Al, 94% using Fe_2O_3/Al, and 87% using Fe_3O_4/Al, respectively. This result may suggest that galvanic couples composed of ZVAl as electron donors and the electron mediators (AC, Fe_2O_3, and Fe_3O_4) are remarkably effective in recovering Au from ammonia thiosulfate solutions. A selective recovery of Cu was observed, and the Cu recoveries were 95% using AC/Al, 21% using Fe_2O_3/Al, and 15% using Fe_3O_4/Al. These results indicate that Fe_2O_3/Al and Fe_3O_4/Al could be used for the selective recovery of gold from copper–ammoniacal thiosulfate solutions.

To investigate the associations of cemented Au and Cu, a morphological analysis of the representative particles on the residues from the cementation experiments was examined in detail using SEM-EDX analysis, as shown in Figures 5–7. The results shown in Figure 5 for the residue using AC/Al were consistent with the previously reported results [16–18]. In the SEM-BSE image, bright (white) domains (points AC-2 and AC-3) were observed on the surface of the grey particle (point AC-1). The point spectrum data showed that, at the grey particles (point AC-1), firm Al and O peaks were detected with a minor S peak, suggesting that the grey particle was ZVAl with an oxidized surface. In the spectrum at the bright domains (AC-2 and AC-3), the Au and Cu peaks were observed together with the C peak, suggesting that the bright domains corresponded to the Au and Cu deposited on the AC.

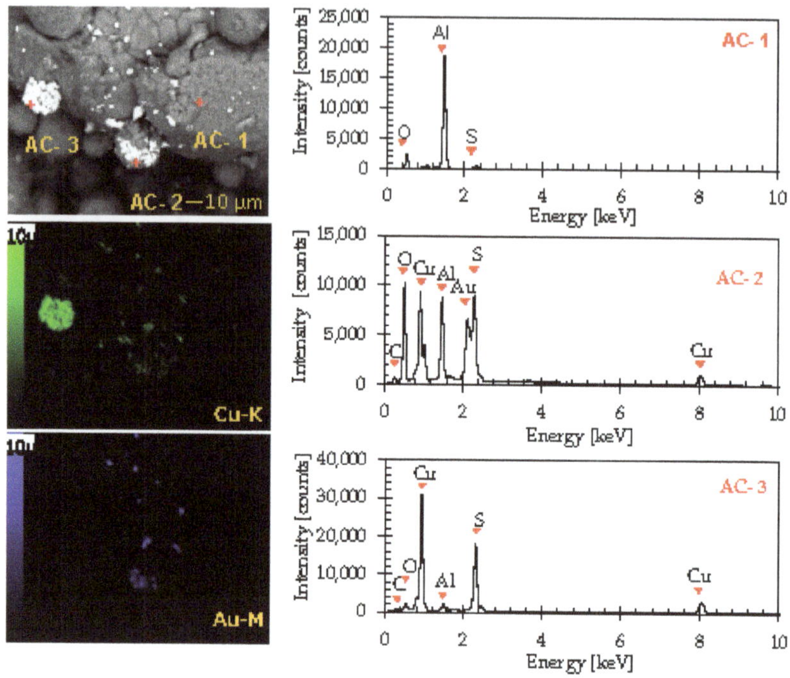

Figure 5. Backscattered electron photomicrograph using an energy-dispersive point analysis with elemental mapping of the leach residue using Al and AC galvanic systems.

The SEM-EDX mapping image confirmed that Au and Cu were always deposited on the AC attached to ZVAl. Figures 6 and 7 show the results of the SEM-EDX analysis for the solid residues obtained in the cementation experiments using Fe_2O_3/Al and Fe_3O_4/Al, respectively. The point spectrum and mapping results shown in Figures 6 and 7 suggested that Au was deposited on the iron oxide particles attached to the ZVAl surface. Noticeably, the copper signal intensity ranged from 200 to 6000 cps, as illustrated in Figure 6 AH-3 and Figure 7 AM-3, while the values were higher in the results (1000–31,000 cps) illustrated in Figure 4 AC-2 and AC-3, implying that the copper deposition was limited for the iron oxide/Al system. An analysis of the spectra from 40 different points on illuminated bright particles was conducted. The results showed that the ratio of the gold signal to copper signal was higher for the residues using iron oxide/Al than those using AC/Al. This result agreed with the result illustrated in Figure 4 and confirmed that Au was preferably cemented compared to Cu when iron oxides were used as the electron mediators.

3.2. Electrochemical Properties of the Fe_2O_3/Al and Fe_3O_4/Al Electrodes

The cementation of Au and Cu from ammonium thiosulfate solutions was an electrochemical process. When ZVAl was used as an electron donor, the following half-cell reaction of an anode reaction was as follows.

$$Al = Al^{3+} + 3e^{-} \tag{4}$$

For the cathodic reactions, the deposition of Cu and Au were assumed to be the following.

$$Cu(NH_3)_4^{2+} + 2e^- = Cu + 4NH_3 \quad (5)$$

$$Au(S_2O_3)_2^{3-} + e^- = Au + 2S_2O_3^{2-} \quad (6)$$

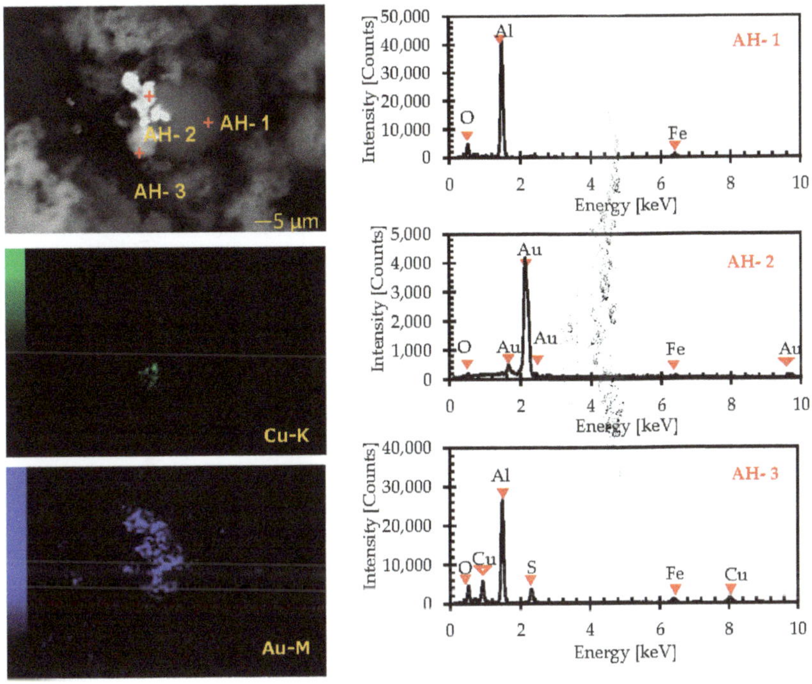

Figure 6. Backscattered electron photomicrograph using an energy-dispersive point analysis with elemental mapping of the leach residue using Al and hematite galvanic systems.

It is important to note that the anodic reaction was typical for the Cu and Au deposition, while the cathodic reactions differed for Cu and Au. This result suggested that the selective deposition of Au observed in the previous sections was because an Au cathodic reaction was preferred to a Cu reaction when iron oxide was used as an electron mediator. A series of electrochemical experiments were conducted to confirm the results.

To identify the reduction potential of $Cu(NH_3)_4^{2+}$ and $Au(S_2O_3)_2^{3-}$, cyclic voltammetry was conducted using a Pt working electrode in NH_4^+/NH_3 buffer solutions containing different electrolytes. Figure 8a presents the effects of $S_2O_3^{2-}$ and the gold ions on the electrochemical experiments. In the voltammogram spectra generated using a control (without metal ions) NH_4/NH_3 buffer solution, the current density was almost zero between −0.7 V and +1.0 V, indicating no electrochemical reaction in the range. However, below −0.7 V, the current decreased with a decreasing electrode potential. This may have been due to a reduction in H_2O to form H_2 gas.

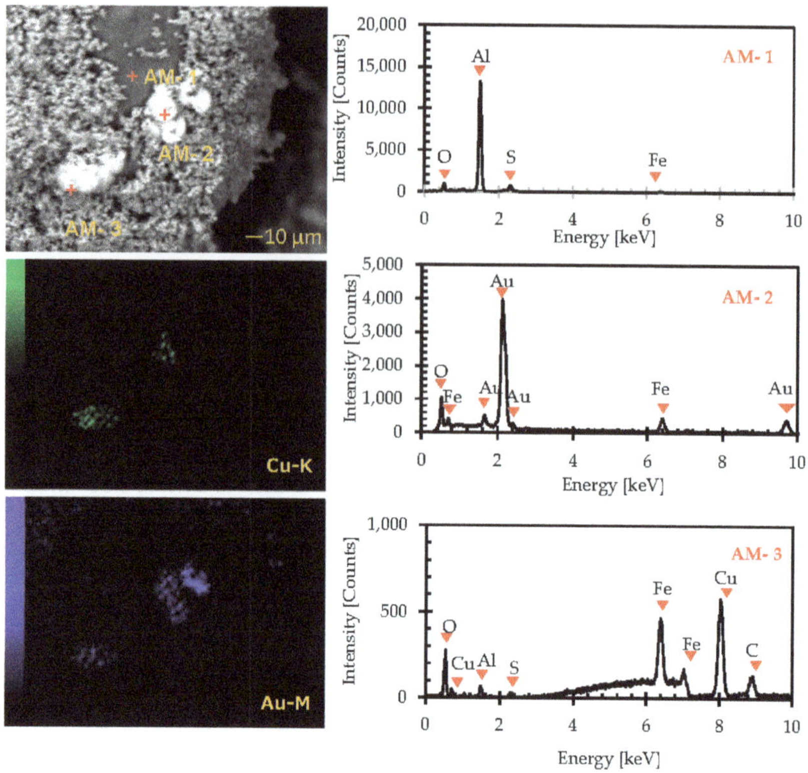

Figure 7. Backscattered electron photomicrograph using an energy-dispersive point analysis with elemental mapping of the leach residue using Al and Magnetite galvanic system.

In the case of $S_2O_3{}^{2-}$, a new cathodic current peak appeared at −0.9 V, which was assigned to the reduction in $S_2O_3{}^{2-}$. In the anodic region in the cyclic voltammogram with $S_2O_3{}^{2-}$, a significant anodic peak was observed around +0.6 V. This peak was observed even when the initial potential scan direction was positive, indicating that the anodic current peak was due to the oxidation of $S_2O_3{}^{2-}$.

When the gold ions were added to $S_2O_3{}^{2-}$, $Au(S_2O_3)_2{}^{3-}$ was formed. In the experimental results with $Au(S_2O_3)_2{}^{3-}$, a similar cathodic current peak was observed at −0.9 V. When comparing the cyclic voltammograms, the peak current density was more substantial with $Au(S_2O_3)_2{}^{3-}$ than with $S_2O_3{}^{2-}$, implying that both $S_2O_3{}^{2-}$ and Au^+ in $Au(S_2O_3)_2{}^{3-}$ were reduced at almost similar redox potentials.

Figure 8b shows the effects of the cupric species on the cyclic voltammogram. When $CuSO_4$ was added to the NH_4/NH_3 buffer solution, $Cu(NH_3)_4{}^{2+}$ was formed. With $Cu(NH_3)_4{}^{2+}$, two cathodic peaks were centered at approximately −0.2 V and −0.5 V, and two anodic peaks appeared. The presence of two cathodic peaks implied that Cu^{2+} in $Cu(NH_3)_4{}^{2+}$ underwent a two-step reduction process. The cathodic peak at −0.2 V corresponded to the reduction in Cu^{2+} to Cu^+, while the peak at −0.5 V corresponded to the reduction in Cu^+ to Cu^0. Two anodic peaks in the cyclic voltammogram with $Cu(NH_3)_4{}^{2+}$ may have been the back reactions of the two-step reduction process. The oxidation of Cu^0 to Cu^+ and Cu^+ to Cu^{2+} occurred at −0.2 V and 0 V, respectively.

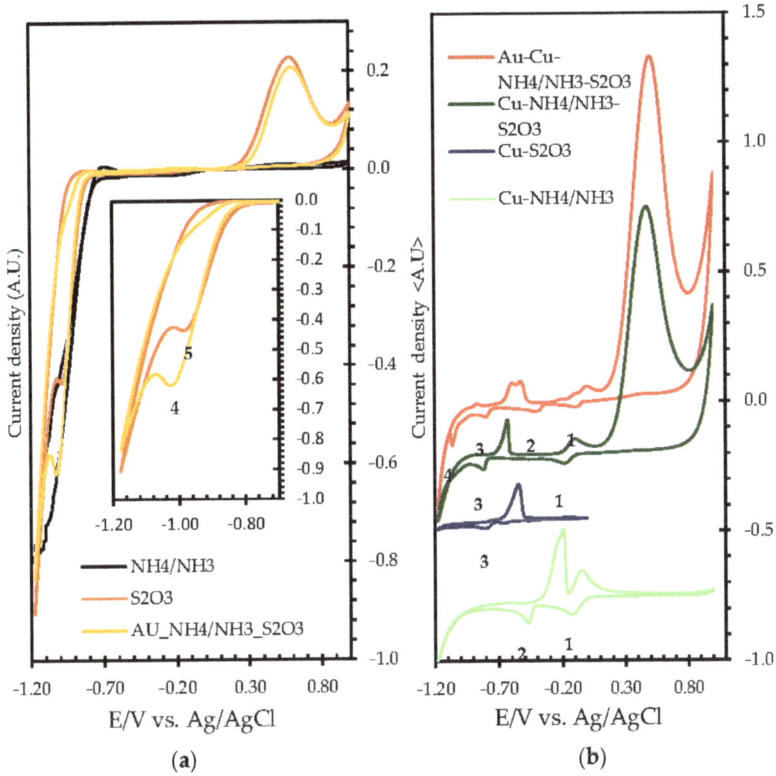

Figure 8. Voltammogram characterization of (**a**) NH_4/NH_3, S_2O_3, and Au-NH_4/NH_3_S_2O_3; and (**b**) Au-CU-NH_4/NH_3-S_2O_3, Cu-NH_4/NH_3, Cu-NH_4/NH_3-S_2O_3 and Cu-S_2O_3.

When $CuSO_4$ and $S_2O_3^{2-}$ were added into the NH_4/NH_3 buffer solution together, two sets of reduction and anodic peaks were observed in the cyclic voltammogram. One was approx. −0.2 V and the other was approx. −0.8 V. The cathodic peak at −0.2 V was also observed in the voltammogram without $S_2O_3^{2-}$, confirming that this peak was due to the reduction in the $Cu(NH_3)_4^{2+}$ to $Cu(NH_3)_2^{+}$ species.

The peak at −0.8 V was not observed without $S_2O_3^{2-}$ and $CuSO_4$, implying that this cathodic peak reduced the cuprous thiosulfate complexes, such as $Cu(S_2O_3)^{-}$ to Cu^0. The cuprous thiosulfate complexes were formed from the amine complex $Cu(NH_3)_2^{+}$, which was the reduction product of Cu $(NH_3)_4^{2+}$. Table 1 summarizes the reduction reactions observed in the cyclic voltammograms, as shown in Figure 8.

Table 1. Summary of the reduction reactions observed in the cyclic voltammograms using Pt as a working electrode, as shown in Figure 8.

Assigned Reductions Reactions	Potential (V)
$Cu(NH_3)_4^{2+} + e^- \rightarrow Cu(NH_3)_2^{+} + 2NH_3$	−0.2
$Cu(NH_3)_4^{+} + e^- \rightarrow Cu^0 + 4NH_3$	−0.5
$Cu(S_2O_3)_2^{3-} + e^- \rightarrow Cu^0 + 2S_2O_3^{2-}$	−0.8
$2S_2O_3^{2-} + 6H_2O + 8e^- \rightarrow 4S^0 + 12OH^-$	−0.9
$Au(S_2O_3)_2^{3-} + 6H_2O + 9e^- \rightarrow Au^0 + 4S^0 + 12OH^-$	−1.0

Figures 9 and 10 illustrate the voltammogram of the Fe_2O_3/Al and Fe_3O_4/Al working electrodes in different electrolyte systems (i.e., 10 mM Cu ions, 10 mM Au ions, 10mM Cu and Au ions, and without metal ions). The cathodic sweep parts were highlighted and projected in the right-side corner of the figures. When comparing the results without metal ions and with only Cu^{2+}, there was no significant difference in the current potential curves, indicating that Cu^{2+} was not reduced on the iron oxide/Al electrodes. On the other hand, the cathodic current with Au^{+} was larger than without metal ions, indicating that the gold ions were reduced on the iron oxide/Al electrodes. The cathodic current with the Cu and Au ions was the same as with the electrolyte containing only Au ions, confirming that only Au was reduced on the electrode.

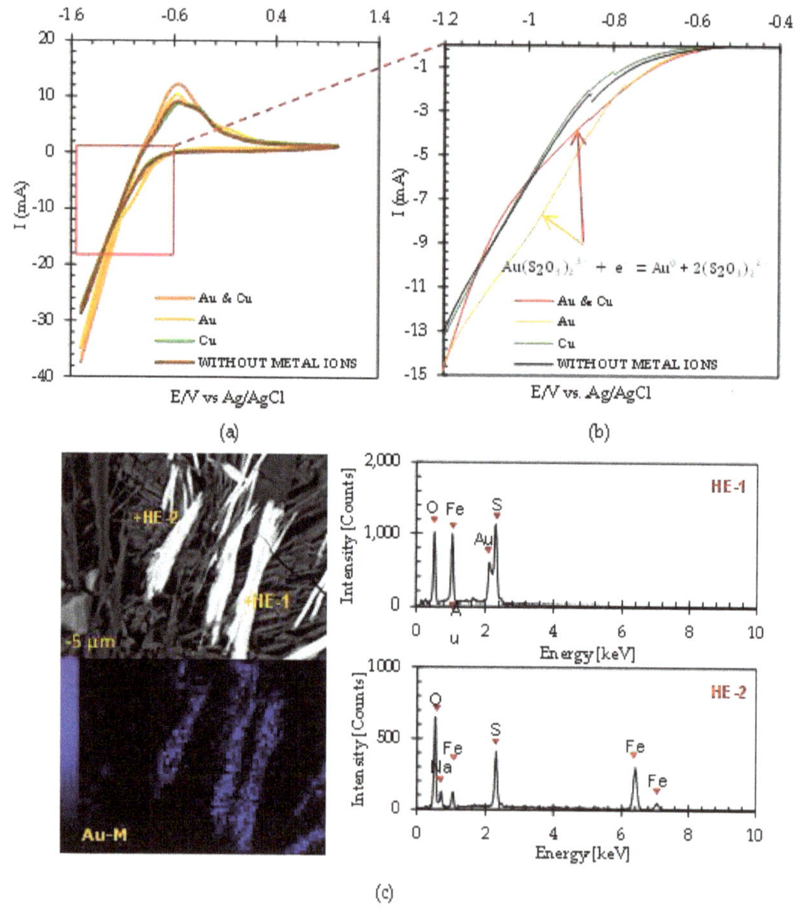

Figure 9. Cyclic voltammogram of the Fe_2O_3/Al electrode at different electrolytes: (**a**) wide scan; (**b**) narrow reduction voltammogram; (**c**) backscattered electron photomicrograph with elemental mapping and a pointed analysis of the Cu–Au electrolytes.

In contrast, Cu^{2+} was not reduced, i.e., the selective reduction of gold ions occurred on the iron oxide/Al electrodes. The previous paper reported that when an Al electrode was used without an electron mediator, there was no significant reduction current due to the deposition of metal ions, such as Cu and Au [17,18]. Therefore, it can be concluded that the reduction in metal ions occurred only when iron oxides were attached to the surface of the Al electrode. This result suggested that the iron oxides functioned as the electron mediators from Al to the metal ions.

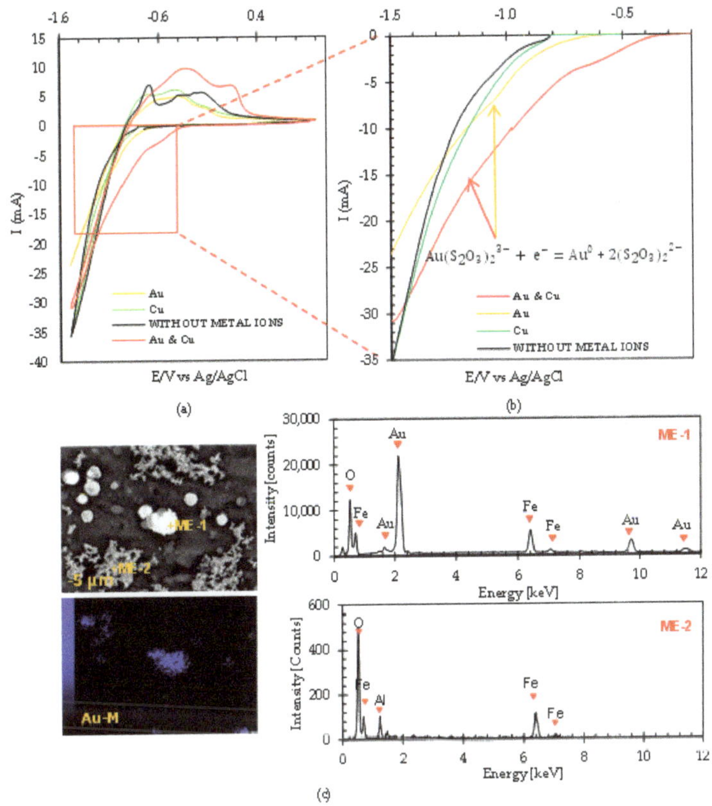

Figure 10. Cyclic voltammogram of the Fe_3O_4/Al electrode at different electrolytes: (**a**) wide scan; (**b**) narrow reduction voltammogram; (**c**) backscatter electron photomicrograph with elemental mapping and a pointed analysis of the Cu–Au electrolytes.

Furthermore, to elucidate the electrochemical behavior of the different improvised electrodes, Figure 11 shows the narrow reduction curves of (a) AC/Al, (b) Fe_3O_4/Al, (c) Fe_2O_3/Al, and (d) the comparative reductive scan using the Au and Cu electrolytes of all improvised electrodes (AC/Al (blue line), Fe_3O_4/Al (red line), Fe_2O_3/Al (yellow line). It can be observed that when using AC/Al, two shoulder-like peaks were observed at approx. −0.5 V and −1.0 V, while for the iron oxides/Al electrode, only one shoulder-like peak was observed and assigned to the reduction in gold ions. Figure 11d elucidates the comparative narrow reductive scan of all improvised electrode. The results suggest that the cyclic voltammogram agreed with the results during the cementation experiments, where only gold ions could be cemented when using iron oxide in both single and binary systems.

Chronoamperometry was conducted by applying a fixed electrode potential of −1.0 V in the stirred solutions for 60 min to confirm the deposition of Au on the working electrodes. After 60 min, the electrodes were taken out and dried to await SEM-EDX analysis. The elemental mapping and point analysis results are illustrated at the bottom side of Figures 9 and 10. The point analysis results showed that Au peaks were observed along with S peaks at the point HE-1, as shown in Figure 9, and the point ME-1, as shown in Figure 10. At the same point, Fe and O peaks were detected, implying that gold ions were reduced on the surface of the iron oxide particles attached to the Al electrode.

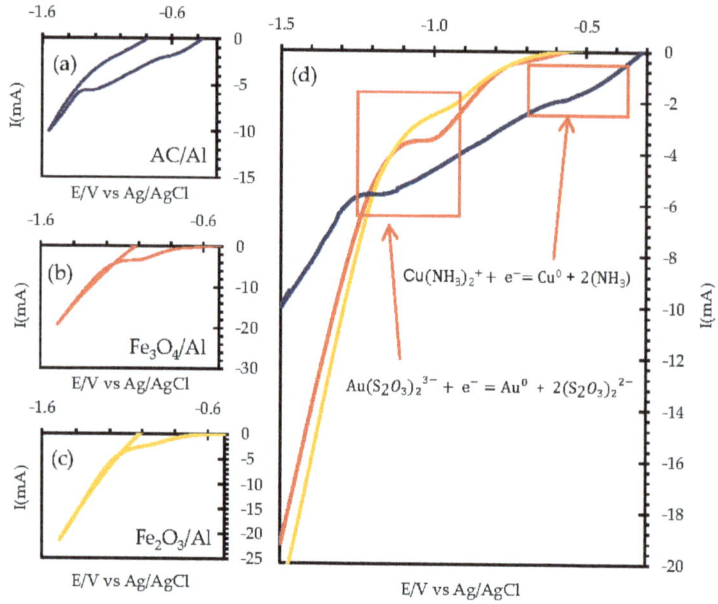

Figure 11. Cyclic voltammogram narrow reduction curve: (**a**) AC/Al, (**b**) Fe_3O_4/Al, (**c**) Fe_2O_3/Al, and (**d**) a narrow reductive scan of all the improvised electrodes at the Cu and Au electrolytes.

As shown in Figure 12, the gold deposited onto the iron oxide showed a similar morphology with the iron oxide particles (morphology, here, pertains to the shape of the iron oxides). The hematite and deposited gold were in longitudinal orientation (Figure 12A) while the magnetite and gold had an endless array of "nodular-like" particles (Figure 12B). These morphological observation results further supported that the iron oxides were an electron mediator between the electron donor (Al) and the electron acceptor, namely the $Au(S_2O_3)_2^{3-}$ ions.

$$Au(S_2O_3)_2^{3-} = Au^+ + 2S_2O_3^{2-} \quad (7)$$

$$2S_2O_3^{2-} + 6H_2O + 8e^- = 4S^0 + 12OH^- \quad (8)$$

$$Au^+ + e^- = Au^0 \quad (9)$$

$$Au(S_2O_3)_2^{3-} + 6H_2O + 9e^- = Au^0 + 4S^0 + 12OH^- \quad (10)$$

3.3. *Proposed Cementation Model*

Figure 13 shows the schematic diagram of the proposed cementation model of gold using iron oxides and ZVAl. In this model, ZVAl acted as an electron donor or anode where the anodic dissolution of ZVAl occurred (Equation (4)). An iron oxide particle attached to the ZVAl particle surface acted as an electron mediator or cathode. Electrons from ZVAl passed through the iron oxide to the gold thiosulfate complexes adsorbed on the surface. The cathodic reaction was the cementation (reductive deposition) of Au and S from $Au(S_2O_3)_2^{3-}$ (Equation (10)). These ions were assigned as electron acceptors. This mechanism was supported by the electrochemical and surface analysis results presented in Figures 9 and 10.

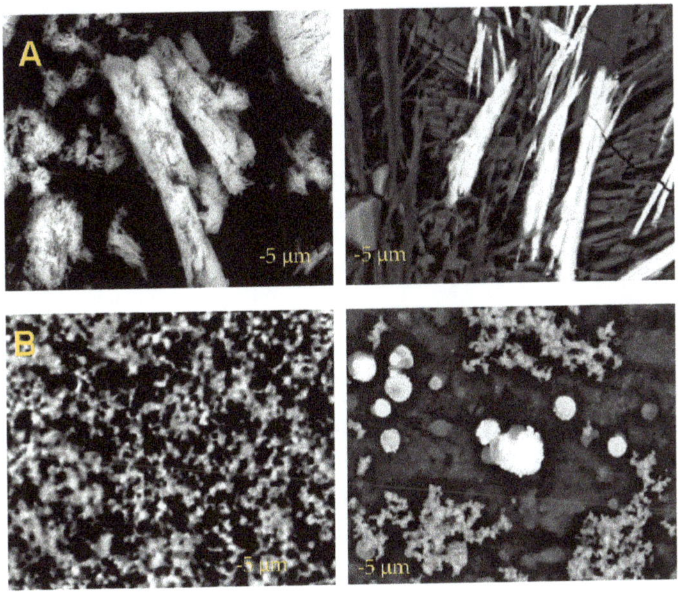

Figure 12. Backscattered electron photomicrograph of (**A**) hematite particles (**left**) and hematite with Au (**right**), and (**B**) magnetite particles (**left**) and magnetite with Au (**right**).

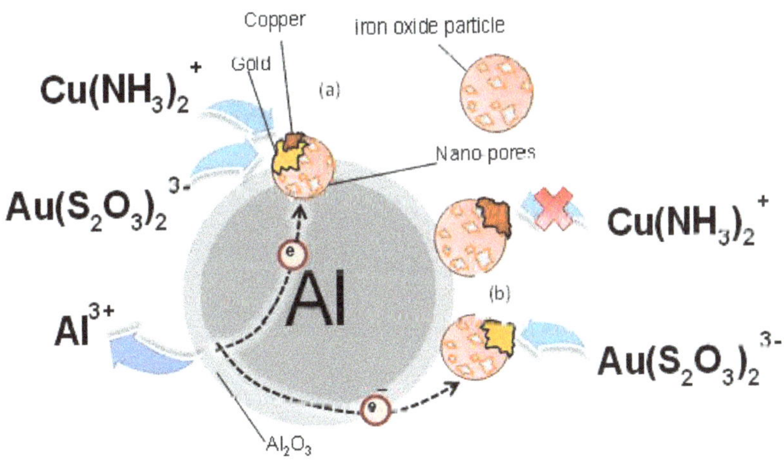

Figure 13. Schematic diagram of the proposed galvanic electron transfer mechanism of iron oxide and zero-valent aluminum. (a) Direct contact on the porosity within the Al–oxyhydroxide layer and (b) direct transfer through the Al–oxyhydroxide layer.

Several interesting matters that can be observed from the proposed model are presented as follows.

First, for the electron transfer mechanism across the Al_2O_3 layer between ZVAl and the iron oxide particle, the ZVAl surface was covered with a fragile layer of Al_2O_3 due to autogenous oxidation. Since Al_2O_3 is an electric insulating material, a direct current could not pass through this layer. The interpretation of the electron transfer through the insulating Al_2O_3 layer was expressed **via** (1) the quantum tunnelling effect [32] and (2) an alternative current transfer through the Al_2O_3 layer acting as a capacitor [16].

The tunnelling effect interpreted the electron transfer when the Al_2O_3 layer was fragile, where a thin Al_2O_3 layer was broken down upon continuous agitation where the particles collided. On the other hand, electron transfer through the Al_2O_3 capacitor was possible when electrochemical reactions (anodic and cathodic reactions) or physical contact with ZVAl and the iron oxides were dynamically changed over time.

Lastly, the most important issue associated with the proposed model was the selective deposition of gold. Gold was deposited on the iron oxide surface when iron oxides were used as an electron mediator. However, the copper deposition was limited (Figures 6 and 7). Furthermore, the reduction potential of $Au(S_2O_3)_2^{3-}$ was much lower than that of $Cu(NH_3)_4^{2+}$, indicating that from the thermodynamic viewpoint, a reductive deposition of copper was more accessible than gold. This result was confirmed when using activated carbon as an electron mediator where both copper and gold were deposited. These results further suggested that the selective gold deposition was due to the properties of the iron oxide used as the electron mediators, and simple thermodynamic considerations could not interpret the mechanism.

The selective deposition of gold in an ammonia thiosulfate leaching system was significant and industrially attractive since this system required cupric ions as a leaching catalyst, and extremely high concentrations of copper ions coexisted with low concentrations of gold leached from the ores. If the selective deposition of gold were possible, separating gold and copper would not require after the deposition step, and the process would become more straightforward compared to the previous research [16–18], where both Cu and Au were cemented using AC as the electron mediator. Thus, understanding the mechanism of selective gold deposition with iron oxide is essential and further studies are needed and may be presented in the following paper.

4. Conclusions

The potential of iron oxides (Fe_2O_3 and Fe_3O_4) for use as an electron mediator and their galvanic interactions with zero-valent aluminum for enhancing gold cementation from a gold–copper ammoniacal thiosulfate solution was evaluated. The cementation experiments confirmed that iron oxides could be a potential candidate as an electron mediator. Different from activated carbon (with a 99% Au and Cu recovery) that was previously proposed as an electron mediator, the selective recovery of gold was obtained at approx. 84–89% Au and the Cu recovery was approx. 15–21% Cu when using iron oxides, even when the ammonia thiosulfate leachate contained a high concentration of copper ions. Furthermore, the electrochemical analysis further validated selective cementation. When using iron oxides, the cyclic voltammograms obtained one "shoulder-like" peak centered at -1.0 V for hematite and magnetite, which was assigned to reducing gold ions. Compared to activated carbon, the voltammogram spectra obtained two "shoulder-like" peaks centered at -1.0 V and -0.5 V, which were assigned to gold and copper reduction, respectively.

Lastly, the surface analysis revealed that the morphology of the metal deposited on the surface of the iron oxide/Al electrode after chronoamperometry mimicked the morphology of the iron oxide particles and was assigned only to gold. With these results, a promising effective technology can be used to treat the century-old problem dealing with pyritic-gold bearing ore, where, after the roasting process, gold can be recovered via a thiosulfate leaching–cementation process without using activated carbon as an absorbent. Though this research was focused on the technical feasibility of iron oxides as an electron mediator, the results warrant an in-depth investigation using natural refractory gold ores, which may be presented in the following paper.

Author Contributions: Conceptualization, J.Z., S.J. and N.H.; formal analysis, J.Z., S.J., M.I. and N.H.; investigation, J.Z., Y.E. and N.H.; methodology, J.Z., S.J., A.K., N.O. and I.P.; project administration, N.H.; resources, N.H.; software, J.Z.; supervision, S.J., I.P., M.I., Y.E. and N.H.; validation, J.Z. and N.H.; writing—original draft, J.Z. and N.H.; writing—review and editing, J.Z., S.J. and N.H. All authors have read and agreed to the published version of the manuscript.

Funding: This research received no external funding.

Data Availability Statement: Supporting data can be accessed upon request to the authors.

Conflicts of Interest: The authors declare no conflict of interest.

References

1. Qin, H.; Guo, X.; Tian, Q.; Zhang, L. Pyrite enhanced chlorination roasting and its efficacy in gold and silver recovery from gold tailing. *Sep. Purif. Technol.* **2020**, *250*, 117168. [CrossRef]
2. Qin, H.; Guo, X.; Tian, Q.; Yu, D.; Zhang, L. Recovery of Gold from sulfide refractory gold ore: Oxidation roasting pretreatment and gold extraction. *Miner. Eng.* **2021**, *164*, 106822. [CrossRef]
3. Konadu, K.T.; Huddy, R.J.; Harrison, S.T.; Osseo-Asare, K.; Sasaki, K. Sequential pretreatment of double refractory gold ore (DRGO) with a thermophilic iron oxidizing archaeon and fungal crude enzymes. *Miner. Eng.* **2019**, *138*, 86–94. [CrossRef]
4. Dong, Z.; Jiang, T.; Xu, B.; Yang, Y.; Li, Q. Recovery of gold from pregnant thiosulfate solution by the resin adsorption technique. *Metals* **2017**, *7*, 555. [CrossRef]
5. Nan, X.-Y.; Cai, X.; Jun, K. Pretreatment Process on Refractory Gold Ores with As. *ISIJ Int.* **2014**, *54*, 543–547. [CrossRef]
6. Xie, F.; Chen, J.-N.; Wang, J.; Wang, W. Review of gold leaching in thiosulfate-based solutions. *Trans. Nonferrous Met. Soc. China* **2021**, *31*, 3506–3529. [CrossRef]
7. Qin, H.; Guo, X.-Y.; Tian, Q.-H.; Zhang, L. Recovery of gold from refractory gold ores: Effect of pyrite on the stability of the thiourea leaching system. *Int. J. Miner. Met. Mater.* **2021**, *28*, 956–964. [CrossRef]
8. Sun, C.-B.; Zhang, X.-L.; Kou, J.; Xing, Y. A review of gold extraction using noncyanide lixiviants: Fundamentals, advancements, and challenges toward alkaline sulfur-containing leaching agents. *Int. J. Miner. Met. Mater.* **2020**, *27*, 417–431. [CrossRef]
9. Tremblay, L.; Deschênes, G.; Ghali, E.; McMullen, J.; Lanouette, M. Gold recovery from a sulphide bearing gold ore by percolation leaching with thiourea. *Int. J. Miner. Process.* **1996**, *48*, 225. [CrossRef]
10. Ma, C.J.; Li, J.Y.; Liu, R.J. A review of thiocyanate hydrometallurgy for the recovery of gold. *Appl. Mech. Mater.* **2015**, *768*, 53. [CrossRef]
11. Ahtiainen, R.; Lundström, M. Cyanide-free gold leaching in exceptionally mild chloride solutions. *J. Clean. Prod.* **2019**, *234*, 9. [CrossRef]
12. Ahtiainen, R.; Lundström, M.; Liipo, J. Preg-robbing verification and prevention in gold chloride-bromide leaching. *Miner. Eng.* **2018**, *128*, 153. [CrossRef]
13. Wang, H. X.; Sun, C.-B.; Li, S.-Y.; Fu, P.-F.; Song, Y.-G.; Li, L., Xie, W. Q. Study on gold concentrate leaching by iodine-iodide. *Int. J. Miner. Met. Mater.* **2013**, *20*, 323–328. [CrossRef]
14. Konyratbekova, S.; Baikonurova, A.; Ussoltseva, G.; Erust, C.; Akcil, A. Thermodynamic and kinetic of iodine-iodide leaching in gold hydrometallurgy. *Trans. Nonferrous Met. Soc. China* **2015**, *25*, 3774. [CrossRef]
15. Jeon, S.; Bright, S.; Park, I.; Tabelin, C.B.; Ito, M.; Hiroyoshi, N. The effects of Coexisting Copper, Iron, Cobalt, nickel and zinc ions on Gold recovery by enhanced cementation via Galvanic Interactions between zero-valent aluminum and activated carbon in ammonium thiosulfate system. *Metals* **2021**, *11*, 1352. [CrossRef]
16. Jeon, S.; Tabelin, C.B.; Takahashi, H.; Park, I.; Ito, M.; Hiroyoshi, N. Enhanced Cementation of gold via galvanic interaction using activated carbon and zero-valent aluminum: A novel Approach to recover gold ions from ammonium thiosulfate medium. *Hydrometallurgy* **2019**, *191*, 105165. [CrossRef]
17. Jeon, S.; Bright, S.; Park, I.; Tabelin, C.B.; Ito, M.; Hiroyoshi, N. A simple and efficient recovery technique for gold ions from ammonium thiosulfate medium by galvanic interactions of zero-valent aluminum and activated carbon: A parametric and mechanistic study of cementation. *Hydrometallurgy* **2021**, *208*, 105815. [CrossRef]
18. Jeon, S.; Bright, S.; Park, I.; Kuze, A.; Ito, M.; Hiroyoshi, N. A kinetic study on enhanced cementation of gold ions by galvanic interactions between aluminum as an electron donor and activated carbon as an electron mediator in ammonium thiosulfate system. *Minerals* **2022**, *12*, 91. [CrossRef]
19. Aylmore, M.; Muir, D. Thiosulfate leaching of gold—A review. *Miner. Eng.* **2001**, *14*, 135. [CrossRef]
20. Arima, H.; Fujita, T.; Yen, W. Gold cementation from ammonium thiosulfate solution by zinc, copper, and aluminum powders. *Mater. Trans.* **2002**, *43*, 485–493. [CrossRef]
21. Xu, B.; Kong, W.; Li, Q.; Yang, Y.; Jiang, T.; Liu, X. A review of thiosulfate leaching of gold: Focus on thiosulfate consumption and gold recovery from pregnant solution. *Met.-Open Access Metall. J.* **2017**, *7*, 222. [CrossRef]
22. Feng, D.; van Deventer, J. Thiosulfate leaching of gold in the presence of pyrite. *Hydrometallurgy* **2006**, *82*, 126–132. [CrossRef]
23. Feng, D.; van Deventer, J. Effect of hematite on thiosulphate leaching of gold. *Int. J. Miner. Process.* **2007**, *82*, 138–147. [CrossRef]
24. Karavasteva, M. Kinetics and deposit morphology of gold cemented on magnesium, aluminum, zinc, iron, and copper from ammonium thiosulfate-ammonia solutions. *Hydrometallurgy* **2010**, *104*, 119–122. [CrossRef]
25. Gallagher, N.P.; Hendrix, J.L.; Milosavljevic, E.G.; Nelson, J.H.; Solujic, L. Affinity of activated carbon towards some gold (I) complexes. *Hydrometallurgy* **1990**, *25*, 305–316. [CrossRef]
26. Choi, S.; Jeon, S.; Park, I.; Ito, M.; Hiroyoshi, N. Addition of Fe_3O_4 as electron mediator for enhanced cementation of cd^{2+} and Zn^{2+} on aluminum powder from sulfate solutions and magnetic separation to concentrate cemented metals from cementation products. *J. Environ. Chem. Eng.* **2021**, *9*, 106699. [CrossRef]

27. Hu, G.; Dam-Johansen, K.; Wedel, S.; Hansen, J.P. Decomposition and Oxidation of Pyrite. *Prog. Energy Combust. Sci.* **2006**, *32*, 295–314. [CrossRef]
28. IAlp; Celep, O.; Paktunç, D.; Thibault, Y. Influence of potassium hydroxide pretreatment on the extraction of gold and silver from a refractory ore. *Hydrometallurgy* **2014**, *146*, 64. [CrossRef]
29. Balarini, J.C.; Polli, L.d.O.; Miranda, T.L.S.; de Castro, R.M.Z.; Salum, A. Importance of roasted sulphide concentrates characterization in the hydrometallurgical extraction of zinc. *Miner. Eng.* **2008**, *21*, 100–110. [CrossRef]
30. Fu, P.; Li, Z.; Feng, J.; Bian, Z. Recovery of Gold and Iron from Cyanide Tailings with a Combined Direct Reduction Roasting and Leaching Process. *Metals* **2018**, *8*, 561. [CrossRef]
31. Tabelin, C.B.; Veerawattananun, S.; Ito, M.; Hiroyoshi, N.; Igarashi, T. Pyrite oxidation in the presence of hematite and alumina: II. Effects on the cathodic and anodic half-cell reactions. *Sci. Total Environ.* **2017**, *581–582*, 126–135. [CrossRef] [PubMed]
32. Ke, B.; Li, Y.; Chen, J.; Zhao, C.; Chen, Y. DFT study on the galvanic interaction between pyrite (100) and galena (100) surfaces. *Appl. Surf. Sci.* **2016**, *367*, 270–276. [CrossRef]

metals

MDPI

Review

Mechanochemical Treatment for the Extraction of Lithium from Hard Rock Minerals: A Comprehensive Review

Yuik Eom [1], Laurence Dyer [1], Aleksandar N. Nikoloski [2] and Richard Diaz Alorro [1,*]

[1] Western Australian School of Mines: Minerals, Energy and Chemical Engineering, Curtin University, Kalgoorlie, WA 6430, Australia; yuik.eom@postgrad.curtin.edu.au (Y.E.); laurence.dyer@curtin.edu.au (L.D.)
[2] College of Science, Technology, Engineering & Mathematics, Murdoch University, 90 South Street, Murdoch, WA 6150, Australia; a.nikoloski@murdoch.edu.au
* Correspondence: richard.alorro@curtin.edu.au

Abstract: Lithium (Li) extraction from Li hard rock minerals involves thermal activation at elevated temperatures and the use of corrosive reagents. The reagents can damage the environment if they are not adequately contained as part of the process, and the high temperatures require large amounts of energy, which can contribute to greenhouse gas emissions unless renewable energy sources are used. These concerns have motivated research into many alternative methods to partially or completely replace the conventional process while maintaining or enhancing the Li extraction rate. Mechanochemistry is one of these alternative methods, as it has the potential to increase the reactivity of Li resources with reagents while reducing the need for corrosive reagents. Mechanochemistry has been applied to mineral processing, particularly introduced as the pre-treatment method before pyro- or hydrometallurgical processes, for several decades. In recent years, a few studies about direct Li extraction from mechanochemically activated hard rock minerals utilizing various co-grinding additives have been reported. This review deals with the application of mechanochemistry to process the Li-bearing hard rock minerals, spodumene and lepidolite, and the mechanisms of the mechanochemical treatment.

Keywords: lithium; hard rock minerals; mechanochemical activation; mechanical activation

1. Introduction

1.1. Lithium Application and Major Lithium Resources

Lithium (Li) is predominantly used in the battery industry, which currently accounts for more than half of the Li end-use market share [1–3]. Li-ion batteries (LIBs) have a high energy density and are therefore, ideal power supply units for portable devices (e.g., cellphones, laptops, and digital cameras) [4,5]. They have been the main battery used for such applications since the early 1990s [4,5]. Li-ion battery technology has advanced over the last three decades, leading simultaneously to product size reduction and increased capacity [4–6]. The usage of Li in LIBs production has continuously increased due to the transition from traditional energy sources to renewable energy, as it is used in energy storage facilities and electric vehicles [5,7]. Moreover, other elements cannot easily be substituted for Li in most applications, so a steady growth of Li consumption is expected [1]. This accelerating growth in the global demand for LIBs has led to renewed expansion of Li mining operations, although it was briefly reversed by a slowdown in 2020 due to the global pandemic [8].

Li is mainly extracted from brine and hard rock deposits. According to the minerals commodity summaries reported by United States Geological Survey (USGS) [8], there are six hard rock mine operations in Australia, two brine operations each in Argentina and Chile, and two brine and three hard rock operations in China currently producing the majority of Li. Unusual sources, such as lacustrine sedimentary basins and clay minerals, with relatively large size reserves, are starting to receive attention but have not yet been widely developed due to the new processing methods required [9,10].

Brine was the dominant Li resource from the 1980s to the 2000s due to its low processing costs, which even resulted in some closures of hard rock mines [11]. The unit production cost from brine processing is around 2300 USD/ton of lithium carbonate (Li_2CO_3), while 6400 USD/ton of Li_2CO_3 for hard rock minerals processing [12,13]. Li extraction from brine generally involves the brine concentration stage, which is done by solar evaporation. This method is very slow and usually takes between 12 and 24 months to obtain final products depending on regional factors such as weather conditions [2,10,14–16]. More advanced technologies such as precipitation, adsorption, and solvent extraction have been investigated to reduce the production period, but the significant differences in the components of each brine source have made it difficult to develop a general production process [16].

Li extraction from hard rock minerals is less susceptible to operational disruption by weather conditions or raw material supply [10]. Therefore, hard rock mining operations can more flexibly respond to market trends than brine-based operations. Recently, the economic importance of hard rock deposits increased due to the high Li price compensating enough for the high processing costs [1,2,10,11,17]. More than 130 Li-bearing minerals have been reported, but only a few (7) are exploited as Li sources and have economic significance [15,18,19]. These minerals are mostly found in pegmatite deposits as a form of Li-aluminosilicate, Li-mica, or fluorophosphate, as described in Table 1. Spodumene ($LiAlSi_2O_6$) has primary economic importance as the most abundant Li-bearing mineral [9,17]. Lepidolite ($K(Li,Al)_3(Si,Al)_4O_{10}(F,OH)_3$) and petalite ($LiAlSi_4O_{10}$), which are usually found together with spodumene, are also important Li resources [3,9,10,20,21].

Table 1. Li-bearing minerals and their Li contents with major mining sites (adapted with permission from Ref. [9]. Copyright 2014 Elsevier and Ref. [17]. Copyright 2020 Elsevier).

Mineral	Formula	% Li Content		Location of Major Mining Site
		Theoretical	Range in Commercial Minerals	
Spodumene [a]	$LiAlSi_2O_6$	3.73	1.9–3.3	Australia
Lepidolite [a]	$K(Li,Al)_3(Si,Al)_4O_{10}(F,OH)_3$	3.56	1.4–1.9	Zimbabwe
Petalite [a]	$LiAlSi_4O_{10}$	2.27	1.6–2.2	Zimbabwe
Amblygonite [a]	$(Li,Na)AlPO_4(F,OH)$	4.74	3.5–4.2	Canada, Brazil
Zinnwaldite [a]	$K(Li,Fe,Al)_3(Si,Al)O_{10}(F,OH)_2$	1.70	1.2–1.3	Czech Republic
Hectorite [b]	$Na_{0.3}(Mg,Li)_3Si_4O_{10}(OH)_2$	0.56	0.4	United State
Jadarite [b]	$LiNaSiB_3O_7(OH)$	3.39	0.1	Serbia

[a] pegmatite, [b] sedimentary rock.

1.2. Conventional Li Extraction Process

The most common process for extracting Li from spodumene requires elevated temperatures from 1000–1100 °C to transform the rigid crystal structure to a porous structure which is more amenable to chemical attack [11,21,22]. Except for the phase transformation stage for spodumene, the subsequent process mainly targets the conversion of Li-bearing minerals into water-soluble compounds, and their extraction routes from different minerals are highly similar [21,23–25]. Salt roasting (e.g., carbonation, sulfation, and chlorination) [26–28] and sulfuric acid (H_2SO_4) roasting/digestion methods [23,25,29,30] are widely investigated, and Li can be recovered as a Li_2CO_3, lithium hydroxide (LiOH), or lithium chloride (LiCl), as shown in Figure 1.

Among the investigated methods, the calcination-H_2SO_4 roasting process has dominated research into Li production, mainly from spodumene, since its first discovery in the 1950s and the first commercial-purpose plant commissioned in 2012 by Galaxy Resources in China [9,11,31]. This conventional thermal treatment greatly consumes energy, and H_2SO_4 digestion/roasting methods generate a significant amount of acid waste [24,32,33]. Thus, it is desirable to improve the existing process and to develop new processes to mitigate the drawbacks, in particularly energy consumption and environmental impact, of the current process [22,32].

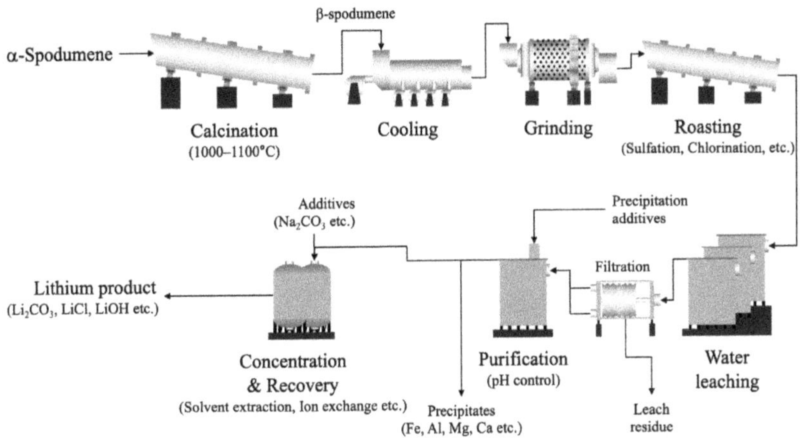

Figure 1. The process flowsheet of the traditional process for Li production from spodumene ores (adapted from Ref. [28]).

The mechanochemical-assisted process has been suggested as one of the alternative approaches to mitigate the operational conditions for Li extraction from spodumene [34–41] and lepidolite [13,37,42–47]. For this purpose, we summarize the previous research into the use of mechanochemical-assisted processes for Li extraction from Li-bearing minerals.

2. Mechanochemistry in General Extractive Metallurgy

Mechanochemistry is a branch of solid-state chemistry that introduces the effect of mechanical energy (e.g., compression, shear, and impact) for the chemical and physico-chemical transformation of materials [24,48–54]. The changes in solid materials by intensive mechanical treatment can be particle size reduction, exposure of fresh interface, creation of active sites, and generation of new bonds for chemical reaction [51,54,55]. These changes are intended to have positive effects on extractive metallurgy by increasing the reactivity of the target material and decreasing the activation energy for achieving the desired reactions [48,56]. As a consequence, the subsequent process flowsheet can be simpler, shorter, less expensive, and produce higher yields [49,51,55,56].

There are two modes for mechanochemical processing: mechanical activation and mechanochemical activation [49,50,57]. In general, mechanical activation enhances the material's reactivity without altering the chemical composition (e.g., amorphization), while mechanochemical activation brings changes in chemical composition through solid-solid reaction (e.g., reduction-oxidation reaction, material synthesis) [13,42,49,50,52]. Hence, mechanical activation does not significantly affect the process flow sheet but usually mitigates operating conditions. In contrast, the intermediate or final product can be produced through mechanochemical activation, and thus, the process flow sheet can be shortened and simplified.

The mechanochemical processing takes place in high-energy milling equipment [49]. Normally, an ultra-fine grinding (UFG) mill, such as a stirring ball mill (attritor), vibration mill, planetary ball mill, or pin mill, is applied [48,56]. In the case of industrial application, several processes incorporated with the mechanochemical process, for example, ACTIVOX^{TM} and MELT processes, employed the stirring mill (e.g., IsaMill, VPX mill, and HIG mill) [49,57,58].

3. Mechanochemical Processing of Li Extraction from Li-Bearing Minerals

Mechanochemical processing in Li extraction has been studied for major hard rock resources (spodumene and lepidolite), and it has been shown to improve the subsequent pyro- and hydrometallurgical processes. The summary of studies on mechanochemical processing is listed in Table 2.

Table 2. A summary of conditions of mechanochemical-assisted process for Li extraction from Li-bearing minerals.

Process	Mineral	Experimental Conditions			Li Leaching Efficiency	Ref.
		Mechanical/Mechanochemical Activation	Thermal Treatment	Leaching		
Calcination	α-SP [a]	Planetary ball mill, 10–30 min	>900 °C (phase transition to β-SP)	N/A [c]	N/A	[34]
Roasting and leaching	α-SP	Planetary ball mill (PMQW series), 2:1 of SP/Na_2SO_4 mass ratio, 20:1 of ball/powder mass ratio, 600 RPM, 5 h	1000 °C, 1 h	H_2O, 40 g/L slurry density, 80 °C, 1 h	<92%	[38]
	β-SP	Ball mill, 2:1 of SP/Na_2CO_3 molar mass ratio, 56:1 of ball/powder mass ratio, 400 RPM, 1 h	400 °C, 10 h	H_2O, 60 °C	86% (Li_2CO_3 recovery yield)	[39]
	LP [b]	Planetary ball mill (PMQW series), 1:1 of LP/Na_2SO_4 mass ratio, 20:1 of ball/powder mass ratio, 600 RPM, 5 h	800 °C, 1 h	H_2O, 40 g/L slurry density, 80 °C, 1 h	>99%	[47]
Acid digestion	LP	Disc mill (N.V. Tema), 5 g/batch, 30 min	98% H_2SO_4 (650 g acid/kg ore), 165 °C, 4 h	H_2O, 1:10 of S/L, 80 °C, 4 h	87%	[44]
		Disc mill (N.V. Tema), 200 g/batch, 15 min	98% H_2SO_4 (1.3 g acid/g ore), 190 °C	H_2O, 1:10 of S/L, 80 °C, 4 h	>90%	[45]
Direct leaching	α-SP	Disc mill (HSM 100), 30 g/batch, 50 min	N/A	0.5 mol/L H_2SO_4, 60 °C	<40%	[36]
		Planetary ball mill, 10–30 min	N/A	20% H_2SO_4, 10 min	96–99% (Al, Li, K, Na)	[34]
		Planetary ball mill (S100), 1:5 of SP/NaF mass ratio, 5–64:1 of ball/powder mass ratio, 500 RPM, 10 h	N/A	Diluted H_2SO_4 (in situ HF generation), 400 RPM, 90 °C, 4 h	<81%	[40]
	β-SP	Planetary ball mill, 50:1 of ball/powder mass ratio, with/without CaO + H_2O, 1000 RPM, 20 min	N/A	H_2O (in situ $Ca(OH)_2$ generation), 95 °C	>90%	[41]
	LP	Planetary ball mill (pulverisette-7), 1:3 of LP/$CaSO_4 \cdot 1/2H_2O$ mass ratio, 700 RPM, 12 h	N/A	H_2O, 10 g/L slurry density, 400 RPM, 25 °C, 1 h	93.5%	[13]
		Planetary ball mill (pulverisette-7), 1:3 of LP/Na_2S mass ratio, 700 RPM, 12 h	N/A	H_2O, 20 g/L slurry density, 25 °C, 0.5 h	93%	[42]

[a] spodumene, [b] lepidolite, [c] not applicable.

3.1. Spodumene

3.1.1. Thermal Treatment

Phase Transformation

In a number of studies, mechanical activation of spodumene has been shown to lower the required calcination temperature for phase transformation [34,35,41,53,59]. Kotsupalo et al. [34] compared the required temperature for phase modification from un-activated and activated α-spodumene to β-spodumene. The formation of β-spodumene was started when the temperature reached 900 °C for mechanically activated α-spodumene, whereas the transition to β-spodumene occurred at 950–1100 °C for un-activated α-spodumene, as shown in Figure 2.

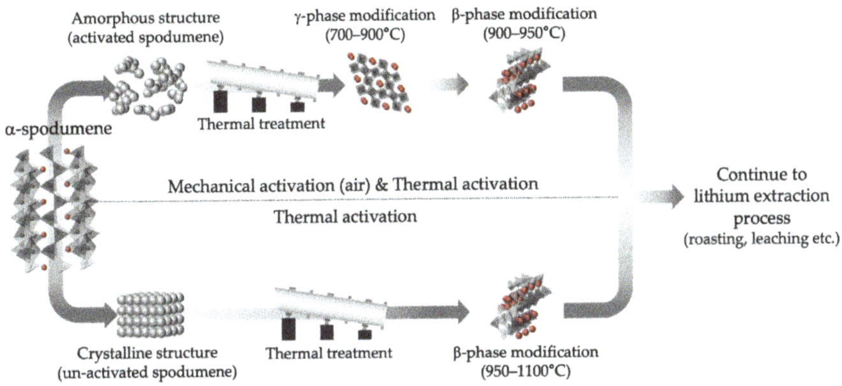

Figure 2. The comparison of required calcination temperature for un-activated and mechanically activated α-spodumene (adapted with permission from Ref. [34]. Copyright 2010 Elsevier).

The main reason for lowering the temperature for phase transformation is closely related to the formation of γ-phase spodumene, which is another allotrope of spodumene that starts to form in the temperature range 700–900 °C [34–36,41]. The damage in chemical bonds (Li/Al-O and Si-O in α-spodumene) and the formation of the amorphous phase during the mechanical activation process results in the formation of γ-spodumene as an intermediate phase in the following thermal treatment. Apart from the formation of the amorphous phase, Gasalla et al. [36] reported that the emergence of the new surface promotes the formation of γ-spodumene as it provides additional nucleation sites. γ-spodumene is stable at higher temperatures and pressures, but it is unstable at ambient pressure [35,41,53,59]. In contrast, β-spodumene is the most thermodynamically stable phase above 900–950 °C under atmospheric pressure, and γ-spodumene starts to convert into β-spodumene at temperatures above 900 °C [60].

Sulfation Roasting

Due to the high-energy consumption and complex purification process of the conventional decrepitation (phase transformation) and acid roasting process, alternative processes involving one-step roasting with alkali sulfates (e.g., sodium sulfate (Na_2SO_4) and potassium sulfate (K_2SO_4)) followed by water (H_2O) leaching have been proposed [11,61,62]. The one-step roasting process enables the direct formation of water-soluble Li salts, which simplifies the process flowsheet and provides more efficient energy consumption [37,62].

Setoudeh et al. [37] used mechanically activated α-spodumene as a starting material for a sulfation roasting process, aiming to enhance the transformation rate and leaching efficiency. The mixture of α-spodumene and Na_2SO_4 was ground for 5 h in a planetary ball mill. However, even after 5 h of mechanical activation time, they could clearly detect the separate α-spodumene and Na_2SO_4 phases through the X-ray diffraction (XRD) analysis, meaning that there was no significant mechanochemically derived reaction between the two materials. Additionally, after 5 h of intensive grinding, there were no discernible changes in the results of sulfation roasting at 800 °C for 1 h.

In a further investigation, Setoudeh et al. [38] introduced two different grinding media, zirconia and stainless steel, to determine the effect of the grinding media on the formation of compounds and the effect of mechanically activated composites on the succeeding roasting process. They leached roasting product in water at 80 °C with 40 g/L slurry density. The type of grinding media had no significant effect on neither the efficiency of the mechanical activation-roasting process nor the synthesis of new materials but only made difference in physical properties (particle size reduction) and mixing efficiency of the feed and reactant. They achieved approximately 92% Li leaching efficiency regardless of grinding media under optimal grinding conditions of a spodumene/Na_2SO_4 mass ratio of 1:0.5, a ball/powder mass ratio of 20:1, 5 h of grinding, and roasting temperature at 1000 °C for 1 h.

Carbonation Roasting

Thermal treatment of β-spodumene using sodium carbonate (Na_2CO_3) to produce Li_2CO_3 directly has been suggested [26,27]. Compared to sulfation roasting and acid/alkaline digestion, the direct carbonation process is able to avoid complex purification processes, reduce processing time, and minimize waste streams [26,39].

Grasso et al. [39] studied the solid-state reaction of the mixture of β-spodumene and Na_2CO_3 prepared by mechanical treatment. Grasso et al. [39] fixed the grinding parameters as follows: a molar ratio of spodumene/Na_2CO_3 of 2:1, a ball/powder mass ratio of 56:1, 400 RPM of grinding speed, and 1 h of grinding time using a planetary ball mill. From the XRD analysis, the evidence of the synthesize of new material between β-spodumene and Na_2CO_3 was not found, but they could obtain 61 wt% and 86 wt% of Li_2CO_3 formation yield after 5 h and 10 h of roasting at 400 °C. Then, they recovered Li_2CO_3 and dehydrated analcime ($NaAlSi_2O_6$), which is a useful absorbent material for contaminant removal, in the following H_2O leaching–filtering–drying process. In this study, Grasso et al. [39] proposed reactions, as described in Equations (1) and (2).

$$2\beta LiAlSi_2O_6(s) + Na_2CO_3(s) = Li_2CO_3(s) + 2NaAlSi_2O_6(s) \quad (1)$$

$$2\beta LiAlSi_2O_6(s) + Na_2CO_3 \cdot 10H_2O(s) = Li_2CO_3(s) + 2NaAlSi_2O_6(s) + 10H_2O(g) \quad (2)$$

dos Santos et al. [27] reported that a higher temperature of above 600 °C is required for maximum conversion to Li_2CO_3 (approximately 71% Li yield) and $NaAlSi_2O_6$ from a β-spodumene-Na_2CO_3-sodium chloride (NaCl) system. On the other hand, Grasso et al. [39] observed the formation of Li_2SiO_3 generated by the reaction between SiO_2 with Li compounds (e.g., Li_2CO_3) at temperatures above 600 °C, which is not desirable for maximizing Li_2CO_3 recovery rate. From the test conditions and results comparison, Grasso et al. [39] noted that the introduction of mechanical treatment could increase reactivity between solid particles by increasing the contact rate and specific surface area and eventually reduce the reaction time and temperature in the following heating process.

3.1.2. Direct Leaching

H_2SO_4 Leaching

The dissolution of β-spodumene in H_2SO_4 is thermochemically favorable and more reactive in acid, as described in Equation (3) [39,63].

$$2\beta LiAlSi_2O_6(s) + H_2SO_4(l) = Li_2SO_4(aq) + 4SiO_2(s) + Al_2O_3(s) + H_2O(l) \quad (3)$$

The high Li recovery from β-spodumene using H_2SO_4 is practically impossible due to the formation of hydrated SiO_2 film on the particle surface [19,64]. Thus, many direct leaching processes require using very strong acid (e.g., hydrofluoric acid (HF)), which is capable of decomposing the Si-O bonds in silicate minerals [65]

On the other hand, H_2SO_4 leaching of α-spodumene without pre-thermal treatment for phase transformation to β-spodumene is not feasible or has very low Li leaching efficiency. This is because α-spodumene has inherently low reactivity with acid or alkaline solutions [17,19]. Alternatively, an autoclave is used to perform high-temperature and high-pressure leaching to increase the reactivity of α-spodumene. There have been several attempts to extract Li directly from mechanically activated α-spodumene [34,36,64]. Gasalla et al. [36] and Kotsupalo et al. [34] reported on the practicality of direct Li leaching from mechanically activated α-spodumene using H_2SO_4.

Gasalla et al. [36] reported that the reactivity of α-spodumene with H_2SO_4 was progressively enhanced as it lost crystallinity as a result of mechanical activation. Gasalla et al. [36] noted that further particle size reduction did not occur after 10 min of grinding time, but α-spodumene continuously lost its original shape and crystallinity while generating superficial porosity. After 50 min of mechanical activation, α-spodumene was leached in 0.5 mol/L H_2SO_4 at 60 °C, and slightly less than 40% of Li and 45% of Al leaching efficiencies were achieved. On the other hand, Kotsupalo et al. [34] leached mechanically activated α-spodumene using higher H_2SO_4 concentrations, 10% and 20%, and reported the alkali metals (Li, K, and Na) and Al almost completely dissolved (96–99% of leaching efficiency). Boldyrev [53] also reported the complete decomposition of alkali metals due to mechanical activation, but this study does not elaborate on the specific methodology for mechanical activation and leaching processes.

The formation of hydrated silicon dioxide (hydro-silica gel) has been reported during the leaching process of mechanically activated α-spodumene in acidic lixiviants [34,36]. Kotsupalo et al. [34] obtained data using immersion microscopy and derivatography to investigate the formation of hydro-silica gel. They reported that the quantity of hydro-silica gel increased from 5% to 60% when mechanical activation time increased from 3 min to 30 min (wet grinding). This hydro-silica gel formation on the surface of the particles could result in low Li recovery, as it hinders the penetration of the lixiviant into the particles [19,64].

HF Leaching

Rosales et al. [40] reported a new direct leaching process using the mechanical activation of the low-grade α-spodumene and sodium fluoride (NaF) mixture. They ground the mixture at a spodumene/NaF mass ratio of 5:1, which is a 1:1 Li/NaF equivalent molar ratio, at 500 RPM of grinding speed for 15–600 min in a planetary ball mill. They noted that NaF acted as a soft matrix in the grinding process, eventually promoting the ionic exchange capacity of the surface of the mineral and, hence, the dissolution of Li in the subsequent leaching process via the reaction Equation (4).

$$\alpha LiAlSi_2O_6(s) + 19HF(l) = LiF(l) + 2H_2SiF_6(l) + H_3AlF_6(l) + 6H_2O(l) \quad (4)$$

From the scanning electron microscope (SEM) images, they observed that NaF undergoes a cold welding process on the particle surface by mechanical action and remains on the surface, subsequently generating HF by reacting with H_2SO_4 in the leaching process (Equation (5)).

$$H_2SO_4(l) + 2NaF(s) = Na_2SO_4(l) + 2HF(l) \quad (5)$$

In the leaching process, H_2SO_4 is not directly involved in Li extraction, but H_2SO_4 can increase the solubility of LiF, while Na^+ ions precipitate hexafluorosilicic acid (H_2SiF_6) as sodium fluorosilicate (Na_2SiF_6) and hydrofluo-aluminic acid (H_3AlF_6) as cryolite (Na_3AlF_6), as described in Equation (6).

$$H_2SiF_6(l) + H_3AlF_6(l) + 5Na^+(aq) = Na_2SiF_6(s) + Na_3AlF_6(s) + 5H^+(aq) \quad (6)$$

Alkaline Leaching

Mcintosh [66] patented an alkaline leaching process for β-spodumene in an autoclave, in which calcium oxide (CaO) is used as the main Li leaching reagent. In this process, CaO and H_2O form milk lime ($Ca(OH)_2$) and then react with β-spodumene at 100–200 °C and 1–17 atm pressure. The overall reaction is described in Equation (7).

$$\beta LiAlSi_2O_6(s) + CaO(s) + H_2O(l) = LiOH(aq) + CaO{\cdot}AlSi_2O_6(s) + H^+(aq) \quad (7)$$

Berger et al. [41] reported that mechanically activated β-spodumene could be processed at atmospheric pressure and below 95 °C, which are milder leaching conditions than Mcintosh reported [66]. Berger et al. [41] tested two grinding methods, β-spodumene with CaO in H_2O (mechanochemical leaching) or without CaO in the air (mechanical activation and then leaching). Regardless of the activation conditions, more than 90% of Li leaching efficiency was achieved within 20 min at a ball/powder mass ratio of 50:1 and 1000 RPM of grinding speed.

3.2. *Lepidolite*

3.2.1. Thermal Treatment

Defluorination

Fluorine (F) in lepidolite is not only a toxic element, but it also reduces Li leaching efficiency. As a result, several previous studies have investigated a defluorination process using steam at a temperature above 800 °C for 30–60 min and have achieved a high Li recovery rate of over 90% [13,31,42,44,67–69]. The defluorination mechanism using steam involves the reaction of a proton (H^+) from steam dissociation with F from lepidolite to form HF. Additionally, hydroxyl groups (OH^-) from steam dissociation react with Si-O bonds in lepidolite to form Si-OH. The produced Si-OH continues to react with OH^- to form new phases, including leucite ($KAlSi_2O_6$) and aluminum silicates [19,67]. As an alternative method for the steam-based defluorination process, in situ capture of F by a roasting process has been considered, but this makes lepidolite processing more complicated [69]. On the other hand, Vieceli et al. [44] observed a defluorination effect when lepidolite is mechanically activated. Vieceli et al. [44] assessed the dehydroxylation (decreased number of hydroxyl group) of initial and mechanically activated lepidolite by thermogravimetry at the peak temperature, which is known to be around 900 °C for lepidolite [70], and the degree of weight loss decreased gradually from 1.75 w/w% for initial lepidolite to almost 0 w/w% for from 30 min activated lepidolite [44,47]. From the result, they proposed that lepidolite lost its hydroxyl interlayer groups during the mechanical activation and made lepidolite's interlayer cohesion weaker due to unbalanced electric charges. Thereby, the temperature requirements for defluorination are decreased from 200–800 °C to 200–500 °C. Although mechanically activated lepidolite showed increased dehydroxylation as the activation time increased, the detailed effect of defluorination induced by mechanical activation on Li leaching efficiency has not been investigated.

Acid Digestion

Acid digestion method for lepidolite mineral processing using H_2SO_4 concentrate, first patented by Botton et al. [71], is one of the most extensively studied methods [29,44–46,72]. Botton et al. [71] achieved 92% of Li leaching efficiency under the following conditions: acid digestion temperature at 165 °C and 72% H_2SO_4 concentration for 4 h, and H_2O leaching for 4 h. However, acid digestion requires large amounts of H_2SO_4 with high concentrations (up to 98%), and it yields acid solid waste (8–10 tons per 1 ton of Li_2CO_3 production), which is a serious environmental concern [24,44,68,73].

Vieceli et al. [44] studied the reactivity of mechanically activated lepidolite concentrate in a disc mill with H_2SO_4 and Li leaching efficiency. Under acid digestion conditions of 165 °C and an acid/lepidolite concentrate mass ratio of 650 g/kg followed by H_2O leaching at 80 °C for 4 h, the Li leaching efficiency increased from 82% to 87% when the mechanical treatment time increased from 10 min to 30 min. When the mechanical treatment time was extended to 45 min, the Li leaching efficiency dropped to 77%, mainly due to the agglomeration of ultra-fine particles observed through SEM and particle size analysis.

Vieceli et al. [45] later optimized the acid digestion process by utilizing the factorial design and response surface methodologies. The mechanical activation time (0–15 min), digestion time (15–120 min), digestion temperature (70–200 °C), and acid/concentrate mass ratio (0.32–1.30 g/g) were changed while maintaining H_2O leaching conditions throughout all experiments. From the regression model of the factor's combination, they noted that the strongest effect on Li leaching was the mechanical activation time. They achieved the highest Li leaching efficiency of over 90% from H_2SO_4 digestion conditions of 15 min mechanical activation time using a disc mill and 15 min of acid digestion time at 190 °C with an acid/concentrate mass ratio of 1.30 g/g.

Vieceli et al. [46] suggested two possible mechanisms for the acid digestion of mechanically activated lepidolite concentrate, as described in Equations (8) and (9). They did not consider the effect of F in the following equations because the defluorination effect occurred during the mechanical activation.

$$K_2(Li_3Al_3)(AlSi_3O_{10})_2(OH)_4 + 5.5H_2SO_4 = 1.5Li_2SO_4 + K_2SO_4 + 3Al(SO_4)(OH) + Al_2O_3{\cdot}6SiO_2 + 6H_2 \quad (8)$$

$$K_2(Li_3Al_3)(AlSi_3O_{10})_2(OH)_4 + 7.5H_2SO_4 = 1.5Li_2SO_4 + K_2SO_4 + 5Al(SO_4)(OH) + 6SiO_2 + 7H_2O \quad (9)$$

Sulfation Roasting

Most Li process from lepidolite use one-step thermal processing [19,74]. Sulfation roasting with various additives followed by H_2O leaching has been investigated for additives including Na_2SO_4, potassium sulfate (K_2SO_4), iron sulfate ($FeSO_4$), and sodium sulfide (Na_2S) [30,31,75]. Among the above reagents, Na_2SO_4 is the most commercially relevant roasting reagent, and several prospective processing operations have investigated this route [68,75].

Setoudeh et al. [47] studied the effect of mechanical activation on lepidolite-Na_2SO_4 mixture as a way to enhance and optimize the Na_2SO_4 roasting process. Using a mechanically activated mixture, the phase transformation during the roasting and Li leaching efficiency in the subsequent H_2O leaching at 80 °C. Setoudeh et al. [47] used milling conditions of a lepidolite/Na_2SO_4 mass ratios of 1:0.5 or 1:1, a ball/powder mass ratio of 20:1, and 5 h of grinding time at 600 RPM of grinding speed in a planetary ball mill. Then, the activated mixture was heated at 700–800 °C for 1 h and leached in H_2O at 80 °C for 1 h. From the XRD analysis data, they detected the peaks of $LiKSO_4$, $LiNaSO_4$, and $Li_2NaK(SO_4)_2$ after roasting. These compounds are reported as Li products in addition to LiF, $Li_2O{\cdot}SiO_2$, $Li_2O{\cdot}2SiO_2$, and trace amounts of $LiAlO_2$ and $LiAlSiO_4$ from the reaction between lepidolite and Na_2SO_4 [31,43]. Additionally, Setoudeh et al. [47] observed albite ($NaAlSi_3O_8$) at 700 °C of roasting temperature with stronger and clearer XRD peaks compared to the study by Yan et al. [75]. They suggested that the solid-solid reaction from mechanical milling increased the interfacial area in lepidolite and created an amorphized crystalline structure, which promoted ion diffusion to the reaction surfaces. In addition, the cracks that developed in mechanically activated lepidolite particles enabled the ions to penetrate deep into the particles, thereby lowering the reaction temperature [47,76]. Nevertheless, they did not observe a significant effect on Li leaching efficiency by introducing lepidolite-Na_2SO_4 together in the intensive grinding process. They achieved more than 99% Li leaching efficiency under grinding conditions of a lepidolite/Na_2SO_4 mass ratio of 1:1, roasting temperature of 800 °C.

3.2.2. Direct Leaching

Lepidolite can be directly leached using various mineral acids (e.g., HCl, H_2SO_4, and HF) [72,77–80]. The drawback of direct leaching of lepidolite using mineral acids is dissolution of multiple impurities, including potassium (K), aluminum (Al), rubidium (Rb), and cesium (Cs), or produces insoluble fluorides, which makes Li recovery more difficult [11,25,77–81]. As a result, direct leaching necessitates the careful determination of the optimum acid dosage for metal dissolution to construct a simple downstream purification process and maximize the high Li production yield [11].

Lee [42] studied direct leaching after the mechanochemical activation of the lepidolite-Na_2S mixture. Lee [42] employed an intensive grinding process and then leached in H_2O instead of using an acid or alkaline lixiviant. The effect of mechanochemical activation on Li leaching efficiency was investigated at different lepidolite/Na_2S mass ratios (1:0–1:3) and different grinding times (0–12 h) while maintaining the other grinding and leaching conditions as follows: 4 g of mixture with 7 zirconia balls (15 mm dia.), grinding speed at 700 RPM, and 30 min of leaching time at 25 °C. Lee observed a similar Li leaching efficiency trend at 1:2 and 1:3 of lepidolite/Na_2S mass ratio and achieved slightly higher than 90% of Li leaching efficiency from 6 h of grinding time. Lepidolite activated without Na_2S (1:0 of mix mass ratio) only reached 4.65% of Li leaching efficiency, and they interred that the major impact on Li leaching is mechanochemical reaction between lepidolite and Na_2S under the intensive stress field rather than fine grinding of itself.

From the same research group, Kim et al. [13] conducted a study on the mechanochemical activation of lepidolite with calcium sulfate hemihydrate ($CaSO_4 \cdot 1/2H_2O$, CSH) mixture. They followed the same procedure previously used by Lee [42]. Similar to the study by Lee [42], the mechanical forces not only destroyed the crystalline structure in lepidolite but also synthesized water-soluble alkali sulfates by adding CSH. Kim et al. [13] investigated Li leaching efficiency by varying the lepidolite/CSH mass ratios (1:0–1:4), grinding times (2–12 h), and leaching times (1–6 h) while keeping the other conditions same with the previous study. The mechanochemically activated mixtures showed similar Li leaching efficiency regardless of the mass ratio of lepidolite/CSH (except for 1:0) and reached higher than 80% of Li leaching efficiency after 6 h of grinding time. The highest Li leaching efficiency was 93.5% at lepidolite/CSH mass ratio of 1:3, 12 h of grinding time, and 4 h of leaching time.

To identify the phase transformation as a result of the mechanochemical process, Kim et al. [13] used XRD analysis. However, due to the high intensity of the XRD peaks from calcium sulfate ($CaSO_4$), which is the dehydrated product of CSH formed during the grinding process, they could not detect other newly synthesized compounds from the XRD patterns. Kim et al. [13] assumed the partial synthesis of lithium sulfate (Li_2SO_4) had occurred during the mechanochemical process, as this is known to be the main water-soluble Li compound [9,13], and in-situ-generated Li_2SO_4 attributed to the enhancement of Li leaching efficiency apart from the destruction of the crystalline structure of lepidolite.

4. Conclusions

The mechanochemical-assisted process has been investigated for Li-bearing mineral processing to mitigate the high operating temperatures and use of corrosive reagents in the conventional process. The mechanochemical-assisted process is regarded to have positive effects on Li-bearing mineral's reactivity and thereby improve the subsequent process by:

1. Breaking the long-range order in the crystal structure, which results in easier nucleation during thermal treatment;
2. Shortening the contact distance between target minerals and additives to enable greater formation of alkali metal salts;
3. In situ formation of water-soluble compounds during the grinding;
4. Enhancing the penetration depth of the lixiviant into the mineral structure to increase Li extractability.

However, mechanical or mechanochemical activation has a minor effect on the recovery efficiency when considering the energy consumption, and its role is not clearly elaborated due to the limitation of analysis techniques and equipment. In addition, most applications of the mechanochemical-assisted process to Li-bearing minerals are currently limited to the improvement of the conventional process and the use of commercially exploited reagents/lixiviants.

For the further development of mechanochemical-assisted processes for Li-bearing hard rock minerals processing, the wide range of studies exploiting the unique mechanisms of mechanochemistry (e.g., mechanochemical activation and mechanochemical leaching) should be considered. Besides spodumene and lepidolite, the unusual resources (e.g., jadarite, hectorite, and smectite) are the potentially applicable resources of the mechanochemical-assisted process. Additionally, the shortage of mechanochemical-assisted processes, including the effect of impurities on the back-end process (e.g., purification and production step) and energy consumption, needs to be addressed in future investigations.

Author Contributions: Conceptualization, R.D.A.; investigation, Y.E.; resources, A.N.N. and R.D.A.; writing-original draft preparation, Y.E.; writing-review and editing, L.D., A.N.N. and R.D.A.; visualization, Y.E.; supervision, L.D., A.N.N. and R.D.A. All authors have read and agreed to the published version of the manuscript.

Funding: This research was funded by Future Battery Industry Cooperative Research Centre. FBICRC Project 030: Beneficiation and Chemical Processing of Lithium Minerals.

Data Availability Statement: No new data were created or analyzed in this study.

Acknowledgments: This paper was prepared as part of the Lithium Beneficiation and Chemical Processing of Lithium Minerals project funded by the Future Battery Industries Cooperative Research Centre as part of the Australian Government Cooperative Research Centres Program.

Conflicts of Interest: The authors declare no conflict of interest.

References

1. Martin, G.; Rentsch, L.; Höck, M.; Bertau, M. Lithium market research—Global supply, future demand and price development. *Energy Storage Mater.* **2017**, *6*, 171–179. [CrossRef]
2. Alessia, A.; Alessandro, B.; Maria, V.G.; Carlos, V.A.; Francesca, B. Challenges for sustainable lithium supply: A critical review. *J. Clean. Prod.* **2021**, *300*, 126954. [CrossRef]
3. Salakjani, N.K.; Singh, P.; Nikoloski, A.N. Production of Lithium—A Literature Review Part 1: Pretreatment of Spodumene. *Miner. Process. Extr. Metall. Rev.* **2019**, *41*, 335–348. [CrossRef]
4. Brodd, R.J. Synopsis of the Lithium-Ion Battery Markets. In *Lithium-Ion Batteries: Science and Technologies*; Yoshio, M., Brodd, R.J., Kozawa, A., Eds.; Springer: New York City, NY, USA, 2009; pp. 1–7. [CrossRef]
5. Schipper, F.; Aurbach, D. A brief review: Past, present and future of lithium ion batteries. *Russ. J. Electrochem.* **2016**, *52*, 1095–1121. [CrossRef]
6. Winslow, K.M.; Laux, S.J.; Townsend, T.G. A review on the growing concern and potential management strategies of waste lithium-ion batteries. *Resour. Conserv. Recycl.* **2018**, *129*, 263–277. [CrossRef]
7. Li, M.; Lu, J.; Chen, Z.; Amine, K. 30 Years of Lithium-Ion Batteries. *Adv. Mater.* **2018**, *30*, 1800561. [CrossRef]
8. USGS. *Mineral Commodity Summaries*; US Geological Survey (USGS): Reston, VA, USA, 2022.
9. Meshram, P.; Pandey, B.D.; Mankhand, T.R. Extraction of lithium from primary and secondary sources by pre-treatment, leaching and separation: A comprehensive review. *Hydrometallurgy* **2014**, *150*, 192–208. [CrossRef]
10. Kesler, S.E.; Gruber, P.W.; Medina, P.A.; Keoleian, G.A.; Everson, M.P.; Wallington, T.J. Global lithium resources: Relative importance of pegmatite, brine and other deposits. *Ore Geol. Rev.* **2012**, *48*, 55–69. [CrossRef]
11. Li, H.; Eksteen, J.; Kuang, G. Recovery of lithium from mineral resources: State-of-the art and perspectives—A review. *Hydrometallurgy* **2019**, *189*, 105129. [CrossRef]
12. Chon, U.; Han, G.C.; Kim, K.Y.; Kim, K.H. Current Status of Lithium Resources. *Resour. Recyc.* **2010**, *19*, 3–8.
13. Kim, B.; Kim, S.; Lee, J. Extraction of Lithium from Lepidolite through Intensive Grinding with Calcium Sulfate Hemihydrate Followed by Water Leaching. *Resour. Recycl.* **2017**, *26*, 47–52. [CrossRef]
14. Helvaci, C.; Mordogan, H.; Çolak, M.; Gündogan, I. Presence and Distribution of Lithium in Borate Deposits and Some Recent Lake Waters of West-Central Turkey. *Int. Geol. Rev.* **2004**, *46*, 177–190. [CrossRef]
15. Talens Peiró, L.; Villalba Méndez, G.; Ayres, R.U. Lithium: Sources, Production, Uses, and Recovery Outlook. *JOM* **2013**, *65*, 986–996. [CrossRef]
16. Khalil, A.; Mohammed, S.; Hashaikeh, R.; Hilal, N. Lithium recovery from brine: Recent developments and challenges. *Desalination* **2022**, *528*, 115611. [CrossRef]
17. Karrech, A.; Azadi, M.R.; Elchalakani, M.; Shahin, M.A.; Seibi, A.C. A review on methods for liberating lithium from pegmatities. *Miner. Eng.* **2020**, *145*, 106085. [CrossRef]
18. Naumov, A.V.; Naumova, M.A. Modern state of the world lithium market. *Russ. J. Non-Ferr. Met.* **2010**, *51*, 324–330. [CrossRef]

19. Choubey, P.K.; Kim, M.; Srivastava, R.R.; Lee, J.; Lee, J.Y. Advance review on the exploitation of the prominent energy-storage element: Lithium. Part I: From mineral and brine resources. *Miner. Eng.* **2016**, *89*, 119–137. [CrossRef]
20. Tadesse, B.; Makuei, F.; Albijanic, B.; Dyer, L. The beneficiation of lithium minerals from hard rock ores: A review. *Miner. Eng.* **2019**, *131*, 170–184. [CrossRef]
21. Aylmore, M.G.; Merigot, K.; Rickard, W.D.; Evans, N.J.; McDonald, B.J.; Catovic, E.; Spitalny, P. Assessment of a spodumene ore by advanced analytical and mass spectrometry techniques to determine its amenability to processing for the extraction of lithium. *Miner. Eng.* **2018**, *119*, 137–148. [CrossRef]
22. Salakjani, N.K.; Singh, P.; Nikoloski, A.N. Production of Lithium—A Literature Review. Part 2. Extraction from Spodumene. *Miner. Process. Extr. Metall. Rev.* **2019**, *42*, 268–283. [CrossRef]
23. Sitando, O.; Crouse, P.L. Processing of a Zimbabwean petalite to obtain lithium carbonate. *Int. J. Miner. Process.* **2012**, *102–103*, 45–50. [CrossRef]
24. Xu, C.; De, S.; Balu, A.M.; Ojeda, M.; Luque, R. Mechanochemical synthesis of advanced nanomaterials for catalytic applications. *Chem. Comm.* **2015**, *51*, 6698–6713. [CrossRef] [PubMed]
25. Gao, L.; Wang, H.; Li, J.; Wang, M. Recovery of Lithium from Lepidolite by Sulfuric Acid and Separation of Al/Li by Nanofiltration. *Minerals* **2020**, *10*, 981. [CrossRef]
26. Chen, Y.; Tian, Q.; Chen, B.; Shi, X.; Liao, T. Preparation of lithium carbonate from spodumene by a sodium carbonate autoclave process. *Hydrometallurgy* **2011**, *109*, 43–46. [CrossRef]
27. dos Santos, L.L.; Maribondo do Nascimento, R.; Pergher, S.B.C. Beta-spodumene:Na_2CO_3:NaCl system calcination: A kinetic study of the conversion to lithium salt. *Chem. Eng. Res. Des.* **2019**, *147*, 338–345. [CrossRef]
28. Fosu, A.Y.; Kanari, N.; Vaughan, J.; Chagnes, A. Literature Review and Thermodynamic Modelling of Roasting Processes for Lithium Extraction from Spodumene. *Metals* **2020**, *10*, 1312. [CrossRef]
29. Zhang, X.; Tan, X.; Li, C.; Yi, Y.; Liu, W.; Zhang, L. Energy-efficient and simultaneous extraction of lithium, rubidium and cesium from lepidolite concentrate via sulfuric acid baking and water leaching. *Hydrometallurgy* **2019**, *185*, 244–249. [CrossRef]
30. Hien-Dinh, T.T.; Luong, V.T.; Gieré, R.; Tran, T. Extraction of lithium from lepidolite via iron sulphide roasting and water leaching. *Hydrometallurgy* **2015**, *153*, 154–159. [CrossRef]
31. Luong, V.T.; Kang, D.J.; An, J.W.; Kim, M.J.; Tran, T. Factors affecting the extraction of lithium from lepidolite. *Hydrometallurgy* **2013**, *134–135*, 54–61. [CrossRef]
32. Gao, T.M.; Fan, N.; Chen, W.; Dai, T. Lithium extraction from hard rock lithium ores (spodumene, lepidolite, zinnwaldite, petalite): Technology, resources, environment and cost. *China Geol.* **2023**, *6*, 137–153. [CrossRef]
33. Song, Y.; Zhao, T.; He, L.; Zhao, Z.; Liu, X. A promising approach for directly extracting lithium from α-spodumene by alkaline digestion and precipitation as phosphate. *Hydrometallurgy* **2019**, *189*, 105141. [CrossRef]
34. Kotsupalo, N.P.; Menzheres, L.T.; Ryabtsev, A.D.; Boldyrev, V.V. Mechanical activation of α-spodumene for further processing into lithium compounds. *Theor. Found. Chem. Eng.* **2010**, *44*, 503–507. [CrossRef]
35. Gasalla, H.J.; Pereira, E. Activation-deactivation mechanisms in spodumene samples. *Solid State Ion.* **1990**, *42*, 1–6. [CrossRef]
36. Gasalla, H.J.; Aglietti, E.F.; Lopez, J.M.P.; Pereira, E. Changes in physicochemical properties of α-spodumene by mechanochemical treatment. *Mater. Chem. Phys.* **1987**, *17*, 379–389. [CrossRef]
37. Setoudeh, N.; Nosrati, A.; Welham, N.J. Enhancing lithium leaching by mechanical activation. *Mong. J. Chem.* **2018**, *19*, 44–48. [CrossRef]
38. Setoudeh, N.; Nosrati, A.; Welham, N.J. Phase changes in mechanically activated spodumene-Na_2SO_4 mixtures after isothermal heating. *Miner. Eng.* **2020**, *155*, 106455. [CrossRef]
39. Grasso, M.L.; González, J.A.; Gennari, F.C. Lithium extraction from β-$LiAlSi_2O_6$ using Na_2CO_3 through thermal reaction. *Miner. Eng.* **2022**, *176*, 107349. [CrossRef]
40. Rosales, G.D.; Resentera, A.C.J.; Wuilloud, R.G.; Rodriguez, M.H.; Esquivel, M.R. Optimization of combined mechanical activation-leaching parameters of low-grade α-spodumene/NaF mixture using response surface methodology. *Miner. Eng.* **2022**, *184*, 107633. [CrossRef]
41. Berger, A.; Boldyrev, V.V.; Menzheres, L. Mechanical activation of β-spodumene. *Mater. Chem. Phys.* **1990**, *25*, 339–350. [CrossRef]
42. Lee, J. Extraction of Lithium from Lepidolite Using Mixed Grinding with Sodium Sulfide Followed by Water Leaching. *Minerals* **2015**, *5*, 737–743. [CrossRef]
43. Vieceli, N.; Nogueira, C.A.; Pereira, M.F.C.; Durão, F.O.; Guimarães, C.; Margarido, F. Optimization of lithium extraction from lepidolite by roasting using sodium and calcium sulfates. *Miner. Process. Extr. Metall. Rev.* **2016**, *38*, 62–72. [CrossRef]
44. Vieceli, N.; Nogueira, C.A.; Pereira, M.F.C.; Dias, A.P.S.; Durão, F.O.; Guimarães, C.; Margarido, F. Effects of mechanical activation on lithium extraction from a lepidolite ore concentrate. *Miner. Eng.* **2017**, *102*, 1–14. [CrossRef]
45. Vieceli, N.; Nogueira, C.A.; Pereira, M.F.C.; Durão, F.O.; Guimarães, C.; Margarido, F. Optimization of an innovative approach involving mechanical activation and acid digestion for the extraction of lithium from lepidolite. *Int. J. Miner. Metall. Mater.* **2018**, *25*, 11–19. [CrossRef]
46. Vieceli, N.; Nogueira, C.A.; Pereira, M.F.C.; Durão, F.O.; Guimarães, C.; Margarido, F. Recovery of lithium carbonate by acid digestion and hydrometallurgical processing from mechanically activated lepidolite. *Hydrometallurgy* **2018**, *175*, 1–10. [CrossRef]
47. Setoudeh, N.; Nosrati, A.; Welham, N.J. Lithium recovery from mechanically activated mixtures of lepidolite and sodium sulfate. *Miner. Process. Extr. Metall.* **2019**, *130*, 354–361. [CrossRef]

48. Baláž, P.; Aláčová, A.; Achimovičová, M.; Ficeriová, J.; Godočíková, E. Mechanochemistry in hydrometallurgy of sulphide minerals. *Hydrometallurgy* **2005**, *77*, 9–17. [CrossRef]
49. Baláž, P.; Baláž, M.; Bujňáková, Z. Mechanochemistry in technology: From minerals to nanomaterials and drugs. *Chem. Eng. Technol.* **2014**, *37*, 747–756. [CrossRef]
50. Warris, C.J.; McCormick, P.G. Mechanochemical processing of refractory pyrite. *Miner. Eng.* **1997**, *10*, 1119–1125. [CrossRef]
51. Mateti, S.; Mathesh, M.; Liu, Z.; Tao, T.; Ramireddy, T.; Glushenkov, A.M.; Yang, W.; Chen, Y.I. Mechanochemistry: A force in disguise and conditional effects towards chemical reactions. *Chem. Comm.* **2021**, *57*, 1080–1092. [CrossRef]
52. Baláž, P. (Ed.) Mechanochemistry in Minerals Engineering. In *Mechanochemistry in Nanoscience and Minerals Engineering*; Springer: Berlin/Heidelberg, Germany, 2008; pp. 257–296. [CrossRef]
53. Boldyrev, V.V. Mechanochemistry and mechanical activation of solids. *Russ. Chem.* **1990**, *39*, 2029–2044. [CrossRef]
54. James, S.L.; Adams, C.J.; Bolm, C.; Braga, D.; Collier, P.; Friščić, T.; Grepioni, F.; Harris, K.D.M.; Hyett, G.; Jones, W.; et al. Mechanochemistry: Opportunities for new and cleaner synthesis. *Chem. Soc. Rev.* **2012**, *41*, 413–447. [CrossRef] [PubMed]
55. Baláž, P.; Takacs, L.; Luxová, M.; Godočíková, E.; Ficeriová, J. Mechanochemical processing of sulphidic minerals. *Int. J. Miner. Process.* **2004**, *74*, S365–S371. [CrossRef]
56. Sandström, Å. Mechanochemical Treatment in Metallurgy: An Overview. In Proceedings of the Conference in Mineral Engineering, Luleå, Sweden, 2–3 February 2016.
57. Odebiyi, O.S.; Du, H.; Liu, B.; Wang, S. Sustainability of valuable metals recovery from hazardous industrial solid wastes: The role of mechanical activation. *J. Sustain. Metall.* **2022**, *8*, 1393–1421. [CrossRef]
58. Baláž, P. Mechanical activation in hydrometallurgy. *Int. J. Miner. Process.* **2003**, *72*, 341–354. [CrossRef]
59. Abdullah, A.A.; Oskierski, H.C.; Altarawneh, M.; Senanayake, G.; Lumpkin, G.; Dlugogorski, B.Z. Phase transformation mechanism of spodumene during its calcination. *Miner. Eng.* **2019**, *140*, 105883. [CrossRef]
60. Dessemond, C.; Soucy, G.; Harvey, J.P.; Ouzilleau, P. Phase Transitions in the α–γ–β Spodumene Thermodynamic System and Impact of γ-Spodumene on the Efficiency of Lithium Extraction by Acid Leaching. *Minerals* **2020**, *10*, 519. [CrossRef]
61. Arne, S.K.; Johan, W.S. Method of Recovering Lithium Salts from Lithium-Containing Minerals. U.S. Patent 2,230,167, 28 January 1941.
62. Qiu, S.; Sun, T.; Zhu, Y.; Liu, C.; Yu, J. Direct Preparation of Water-Soluble Lithium Salts from α-Spodumene by Roasting with Different Sulfates. *Ind. Eng. Chem. Res.* **2022**, *62*, 685–697. [CrossRef]
63. Distin, P.A.; Phillips, C.V. The acid extraction of lithium from the granites of South West England. *Hydrometallurgy* **1982**, *9*, 1–14. [CrossRef]
64. Samoilov, V.I.; Borsuk, A.N.; Kulenova, N.A. Industrial methods for the integrated processing of minerals that contain beryllium and lithium. *Metallurgist* **2008**, *52*, 725–730. [CrossRef]
65. Fogler, H.S.; Lund, K.; McCune, C.C. Acidization III—The kinetics of the dissolution of sodium and potassium feldspar in HF/HCl acid mixtures. *Chem. Eng. Sci.* **1975**, *30*, 1325–1332. [CrossRef]
66. Mcintosh, N.C. Production of Lithium Compounds. U.S. Patent 2,413,644, 31 December 1946.
67. Yan, Q.; Li, X.; Yin, Z.; Wang, Z.; Guo, H.; Peng, W.; Hu, Q. A novel process for extracting lithium from lepidolite. *Hydrometallurgy* **2012**, *121–124*, 54–59. [CrossRef]
68. Yan, Q.; Li, X.; Wang, Z.; Wu, X.; Guo, H.; Hu, Q.; Peng, W.; Wang, J. Extraction of valuable metals from lepidolite. *Hydrometallurgy* **2012**, *117–118*, 116–118. [CrossRef]
69. Li, J.; Kong, J.; Zhu, Q.; Li, H. In-situ capturing of fluorine with CaO for accelerated defluorination roasting of lepidolite in a fluidized bed reactor. *Powder Technol.* **2019**, *353*, 498–504. [CrossRef]
70. Schomburg, J.; Zwahr, H. Thermal differential diagnosis of mica mineral group. *J. Therm. Anal.* **1997**, *48*, 135–139. [CrossRef]
71. Botton, R.; Deigrange, J.P.; Esinet, L.; Steinmetz, A. Method of Recovering Lithium from Lepidolite. U.S. Patent 3,189,407, 15 June 1965.
72. Liu, J.; Yin, Z.; Li, X.; Hu, Q.; Liu, W. Recovery of valuable metals from lepidolite by atmosphere leaching and kinetics on dissolution of lithium. *Trans. Nonferr. Met. Soc. China* **2019**, *29*, 641–649. [CrossRef]
73. Su, H.; Ju, J.; Zhang, J.; Yi, A.; Lei, Z.; Wang, L.; Zhu, Z.; Qi, T. Lithium recovery from lepidolite roasted with potassium compounds. *Miner. Eng.* **2020**, *145*, 106087. [CrossRef]
74. Onalbaeva, Z.S.; Samoilov, V.I.; Kulenova, N.A.; Zhakupova, G.B.; Adylkanova, M.A.; Kokaeva, G.A.; Abdulina, S.A. Procedure for lepidolite concentrate processing. *Russ. J. Appl. Chem.* **2016**, *89*, 1728–1730. [CrossRef]
75. Yan, Q.; Li, X.; Wang, Z.; Wu, X.; Wang, J.; Guo, H.; Hu, Q.; Peng, W. Extraction of lithium from lepidolite by sulfation roasting and water leaching. *Int. J. Miner. Process.* **2012**, *110–111*, 1–5. [CrossRef]
76. Welham, N.J. Mechanical activation of the solid-state reaction between Al and TiO_2. *Mater. Sci. Eng. A* **1998**, *255*, 81–89. [CrossRef]
77. Liu, J.; Yin, Z.; Liu, W.; Li, X.; Hu, Q. Treatment of aluminum and fluoride during hydrochloric acid leaching of lepidolite. *Hydrometallurgy* **2020**, *191*, 105222. [CrossRef]
78. Samoilov, V.I.; Onalbaeva, Z.S.; Adylkanova, M.A.; Kokaeva, G.A.; Abdulina, S.A. Complex loosening of lepidolite concentrate by sulfuric acid. *Metallurgist.* **2018**, *62*, 29–33. [CrossRef]
79. Wang, H.D.; Zhou, A.A.; Guo, H.; Lü, M.H.; Yu, H.Z. Kinetics of leaching lithium from lepidolite using mixture of hydrofluoric and sulfuric acid. *J. Cent. South Univ.* **2020**, *27*, 27–36. [CrossRef]

80. Guo, H.; Kuang, G.; Li, H.; Pei, W.; Wang, H. Enhanced lithium leaching from lepidolite in continuous tubular reactor using H_2SO_4+H_2SiF_6 as lixiviant. *Trans. Nonferr. Met. Soc. China* **2021**, *31*, 2165–2173. [CrossRef]
81. Mulwanda, J. Extraction of Lithium from Lepidolite and Spodumene. Doctoral Dissertation, Murdoch University, Perth, Australia, 2021.

metals MDPI

Review

Technospheric Mining of Critical and Strategic Metals from Non-Ferrous Slags

Bona Lim [1], Mark Aylmore [2] and Richard Diaz Alorro [1,*]

[1] Western Australian School of Mines: Minerals, Energy and Chemical Engineering, Curtin University, Bentley, WA 6102, Australia; Bona.Lim@curtin.edu.au
[2] Mineral Resources Ltd., 20 Walters Drive, Osborne Park, WA 6107, Australia; Mark.Aylmore@mrl.com.au
* Correspondence: Richard.Alorro@curtin.edu.au

Abstract: The technosphere consists of material stocks accumulated by human activities, which can include processing residue, such as slag. Various smelting processes generate slag, and some valuable elements are concentrated in this by-product. In this review, the extraction of critical and strategic metals from non-ferrous slags is discussed. Critical and strategic metals are materials that are vital for the nation's economy and defence, as well as its industries, and have common features, such as expected shortfalls, increasing demand, and few substitutions. There are several definitions, methods, and classifications of critical and strategic elements by different organisations. In this study, reports from seven institutions around the world are summarised, and a list of recommended critical and strategic metals is presented. Non-ferrous slags contain a considerable amount of critical and strategic elements, and research on technology and process development using both pyro- and hydrometallurgical methods is very attractive. When it comes to the extraction of values from slag and the development of technology, it is not only important to consider the economic aspect but also to ensure the processes are low in emissions and energy consumption but high in efficiency and recycling.

Keywords: non-ferrous slag; technosphere; technospheric mining; critical and strategic metals; waste utilisation

1. Introduction

The extraction of metals from ores using high-temperature processing, or pyrometallurgy, produces slag as a by product. Slag is a fusible or molten mass formed when flux, on being subjected to heat, reacts with impurities in the ore. Slag floats on top of matte and can be collected separately [1]. It can normally be divided into two groups: ferrous slags and non-ferrous slags [2]. Other slags, such as those generated from municipal waste, are excluded from this study to avoid presenting an overwhelming amount of information. The focus of this review is non-ferrous slag from metallurgical operations.

Slag is regarded as waste and has applications in brick, cement, and construction [1]; however, these is much attention due to the valuable metals it contains. Slag may cause environmental concerns for its release of potentially toxic elements; however, utilising slags as a secondary source will reduce the possibility of environmental exposure [2–7].

While there is abundant information available on copper slag, the characteristics of other slags such as nickel, tin, and vanadium slags are relatively less known, and their considerations as a potential secondary source of valuable metal warrant detailed characterisation and analysis of physical and chemical properties both for metal extraction and environmental purposes [2,7]. Subsequently, the recovery of valuable metals from ferrous and non-ferrous slag sources has been reviewed from the perspective of repurposing and reprocessing the material [6]. Ferrous slags have been effectively utilised for various applications such as cement production, road construction, and others [6]. Non-ferrous slags also have the potential to be utilised in those applications; however, they contain critical and strategic metals, which makes them more valuable as secondary sources for these metals. This review will focus on the contents of critical and strategic elements in non-ferrous slags and the extraction methods to recover them from the slag. Projected shortfalls, increasing demand, and limited substitutions make metals critical and strategic, and it is hard to secure the supply even though these metals are essential.

As mentioned, there are a few general review articles on slag, and these include the following: characterisation [2,7,8], recovery of values from slag [3,6,9], the environmental aspect [2], bioleaching potential [10,11], each commodity such as copper [12], vanadium [13], titanium [14], tungsten [15], and cobalt [4]. There has been no literature published that focuses specifically on the critical and strategic metals, particularly those contained in non-ferrous slags, and on the potential of critical and strategic metal extraction from these materials. In this review paper, the concept of technospheric mining and its application to the extraction of critical and strategic metals from non-ferrous slags are discussed. This paper also details the different critical and strategic elements found in non-ferrous slags and the processes studied and applied to extract them.

2. Research Methodology

This review paper is devoted to exploring the technospheric mining of critical and strategic metals from non-ferrous slags by introducing the terms, reviewing the literature database, and exploring relevant technologies in hydro- and pyrometallurgy.

The references in this paper were searched through 'Google Scholar' using keywords and were carefully chosen to reflect relevance to the topic. Publications on technospheric mining and the technosphere were used to provide a summary of the concept, except for dissertations. Most reports were chosen to introduce the classifications and methodologies of critical and strategic metals from each institute. There are also other countries or organisations available, for example India and Korea, that provide reports on critical and strategic metals. However, only seven geologically representative institutions were chosen for this review.

An extensive literature review was conducted on the recovery of critical and strategic metals from each slag using various methods. In terms of extraction methods, hydrometallurgical and pyrometallurgical methods such as leaching or roasting were the key technologies to focus on over mineral processing and bio-treatment methods unless specifically relevant and useful to restrict the scope. Some slags have limited literature available. On the other hand, copper slag had extensive literature; thus, strict and careful criteria were used to select the relevant literature; for example, the extraction of critical and strategic metals in metallic or semi-metallic forms was the main focus of choosing the literature.

3. Technosphere and Technospheric Mining

The technosphere contains materials, known as technospheric stocks, which are generated and accumulated by human and technological intervention. These stocks may include waste repositories and by-products, such as urban and industrial waste, process residue, mine waste, process effluents, slag, and tailings [16,17]. Researchers from the 1990s introduced, defined, and classified technospheric stocks and their status based on several factors relating to material flow and composition [17–21]. They can also be categorised based on location, type of stock, management, and state of the stocks. The subgroups of technospheric stocks are shown in Table 1 [17,22]. The concept of mining the technosphere refers to technospheric mining.

Technospheric mining is a term and concept referring to recovering minerals or metals from material stockpiles established by human activity and anthropogenic processes [17]. It is an umbrella concept that incorporates various terminologies common in a circular economy, such as resource recycling and urban mining. Johansson et al. [17] suggested a taxonomy in technospheric mining to address a better understanding of this newly developing field, providing a more holistic approach to reprocessing or repurposing unwanted, end-of-life, or waste materials. For example, the taxonomy also described and reviewed landfill mining [23] and urban mining [24]. There are a number of publications now highlighting the application and importance of technospheric mining and the development of technologies to recover valuable metals from technospheric stocks [25–28].

Table 1. Sub-classification of technospheric stocks.

Technospheric Stocks	Status	References
In-use stocks (In-use mining)	Active	[17]
Landfills (Landfill mining)	Controlled inactive	
Tailings (Tailing mining)		
Slags (Slag mining)		
Hibernating stocks (Hibernation mining)	Uncontrolled inactive	
Dissipated stocks (Dissipation mining)		
Urban technosphere	Active	[22]
Rural technosphere	Uncontrolled inactive	
Subterranean technosphere	Controlled inactive	
Marine technosphere	Uncontrolled inactive	
Aerial technosphere	Uncontrolled inactive	

Some of the technospheric stocks, such as tailing, slag, and metallurgical by-products, have drawn attention since they contain a significant amount of critical and strategic metals [28,29]. However, accumulated tailings or slags potentially contain materials that are hazardous for humans and the environment and can cause tragic accidents when managed inappropriately. Thus, employing the technosphere as a secondary source of resources can contribute to mitigating those environmental problems. Therefore, mining the technosphere is a desirable way to reach sustainable development in social and environmental sections, along with obtaining critical and strategic metals from secondary sources.

Due to projected shortfalls, increasing demand, and limited substitutions of critical and strategic metals, the mining industry needs to pay attention to the exploitation of secondary resources that are already extracted and stored as waste to secure critical and strategic metals. Advanced technology, economic and political advocacy, and the enormous amount of technospheric stock support technospheric mining in becoming a new business model. Technospheric mining not only provides secondary sources for critical and strategic metals but also assists in achieving a circular economy that promotes sustainable development, especially in the environment. Overall, technospheric mining can provide economic, environmental, and industrial advantages.

4. Critical and Strategic Metals

Critical metals are essential for the economy and trade, and strategic metals are needed for a nation's defence and its economy [30]. There are multiple definitions, methodologies, and classifications of critical and strategic metals from various institutes [31]. Although several publications on critical and strategic materials or metals have significantly escalated since the 2010s [32], the concept of strategic and critical raw materials publicly appeared in 1939 with a concern for securing those metals that are required in case of war but are unobtainable within the country and thus need to be imported from other countries. The list of critical and strategic materials in 1939 from the United States included antimony, asbestos, bauxite, camphor, cobalt, cryolite, ferrous alloys, iodine, mercury, opium, quinine, platinum, potash, rubber, and tin [33,34], of which some are still critical and strategic in society and others are not anymore.

It is worthwhile to note that criticality is a degree of how critical each element or metal is; for example, from the measure of 0 to 100, rather than the binary state of 'critical' or 'non-critical'. Particularly, critical metals refer to those that have increasing demand with scarce supply and, therefore, become more important in the industry [35]. Similar to critical metals, some metals that are essential for a nation's defence are classified as strategic, and they share distinct features: (1) Strategic metals are concentrated in a few countries, causing resource nationalism: (2) They are highly essential in the high-technology industry, for example, solar panels, wind turbines, and superalloys: (3) They are irreplaceable, which means few replacements for the strategic metals are available [35,36]. Several articles [37,38] mentioned the terms 'strategic' and 'strategic metals' but did not clarify what the terms mean or the methodology.

Critical and strategic metals are getting more attention with the increasing number of publications and reports, yet there is no comprehensive definition that is agreed on by the institutes or governments around the globe since the geological, political, infrastructural, and environmental situations and factors are divergent [31,32]. A conundrum in this situation is whether it is possible to have a globally defined, long-term classification and methodology for critical and strategic metals. As technology and industry evolve, the factors in the methodology change rapidly, which inevitably results in the alteration of critical and strategic metals. Jim et al. [31] also argued that having a comprehensive and concrete methodology would be ideal, although the amount of work for assessment can be incredibly large; thus, simplifying the assessment can reduce the work and, at the same time, the precision of the result as well.

Representatively, national organisations and governments, in alphabetical order, the Australian Government (AG), the British Geological Survey (BGS), the Chinese Government (CG), the European Commission (EC, referred to as EU), the Ministry of Mines from Government of India (GI), the Organisation for Economic Co-operation and Development (OECD), and the United States Department of the Interior from the United States of America (US DoI, referred to as the USA), released publications on critical and strategic materials and their strategies to secure those valuable metals.

Table 2 shows the critical and strategic metals selected by the institutions listed above. X in Table 2 means the institute chose the element as critical and/or strategic. Among metals described by several organisations, those shared in the list in Table 2 are classified as critical and strategic. This table, however, excludes non-metals such as borate (BO_3^{3-} or $B_4O_7^{2-}$), helium (He), fluorine (F), phosphorus (P), and natural rubber [39–41], while acknowledging that the non-metals are vital for many industries. The purpose of Table 2 is to compare the common critical and strategic metals that were chosen from the seven institutes, regardless of the geological factors. For instance, all institutes categorised REEs as critical and strategic, whereas only two institutes consider cadmium (Cd) as critical and strategic. Metals classified as critical and strategic by two institutes out of seven can be considered low priority globally, for example, rhenium (Re), silver (Ag), thorium (Th), uranium (U), as well as Cd. Some metals that were chosen by only one institute were eliminated from Table 2, considering their criticality not to be as globally significant as others. The metals that were chosen only once were the following: caesium (Cs), rubidium (Rb), tellurium (Te) by the USA [42], and mercury (Hg) by BGS [43].

The Australian government initiated the Critical Minerals Strategy, which entails the vision of critical minerals by 2023, and became the global supplier of those minerals. The Critical Minerals List provides not only the critical minerals that are crucial to the technologies necessary for achieving net zero emissions as well as energy security, but also the strategic materials that are regarded as a 'watchlist' to ensure there is no disruption on the supply chain [40]. Among the metals and materials listed in Table 2, the strategic materials classified by Australia are Al, Cu, Ni, P, Sn, and Zn.

The British Geological Survey published a report on the supply risk of chemical elements and its methodology. The supply risk index rates each element from 4.5 to 9.5. The element with the highest index is Rare Earth Element (REE) with a rate of 9.5, and the lowest-rated metal was gold with 4.5, indicating incredibly insignificant risk. The critical metals chosen by the BGS alone are shown in Table 2. A few metals, for example, mercury (Hg) and uranium (U), were excluded from the list because they were chosen by only one institute [43].

Table 2. Classification of critical and strategic metals.

Metals	AG [40]	BGS [43]	CG [41]	EU [39]	GI [44]	OECD [36]	USA [42]
Aluminium (Al)	X	X	X	X	X	X	X
Antimony (Sb)	X	X	X	X			X
Arsenic (As)	X	X		X		X	X
Barium (Ba)		X		X [1]			X
Beryllium (Be)	X	X		X			X
Bismuth (Bi)	X	X		X			X
Cadmium (Cd)		X				X	
Carbon * (C)	X	X	X	X	X	X	X
Chromium (Cr)	X	X	X		X	X	X
Cobalt (Co)	X	X	X	X	X	X	X
Copper (Cu)	X	X	X	X	X	X	
Fluorspar (F)		X	X				X
Gallium (Ga)	X	X		X		X	X
Germanium (Ge)	X	X		X		X	X
Gold (Au)		X	X			X	
Hafnium (Hf)	X			X		X	X
HREEs	X	X	X	X	X	X	X
Indium (In)	X	X			X	X	X
Iron (Fe)		X	X		X	X	
Lead (Pb)		X			X	X	
Lithium (Li)	X	X	X	X	X	X	X
LREEs	X	X	X	X	X	X	X
Magnesium (Mg)	X	X		X	X	X	X
Manganese (Mn)	X	X		X		X	X
Molybdenum (Mo)	X	X	X		X	X	
Nickel (Ni)	X	X	X	X	X	X	X
Niobium (Nb)	X	X		X	X	X	X
PGMs	X	X		X		X	X
Rhenium (Re)	X	X					
Selenium (Se)	X	X				X	
Silicon metal (Si)	X			X	X		
Silver (Ag)		X				X	
Strontium (Sr)		X		X	X		
Tantalum (Ta)	X	X		X			X
Thorium (Th)	X	X					
Tin (Sn)	X	X	X			X	X
Titanium (Ti)	X	X		X	X	X	X
Tungsten (W)	X	X	X	X			X
Uranium (U)		X	X				
Vanadium (V)	X	X		X	X	X	X
Zinc (Zn)	X	X				X	X
Zirconium (Zr)		X	X		X	X	X

HREEs = Heavy Rare Earth Elements	Dysprosium, Erbium, Europium, Gadolinium, Holmium, Lutetium, Terbium, Thulium, Ytterbium, Scandium, Yttrium
LREEs = Light Rare Earth Elements	Cerium, Lanthanum, Neodymium, Praseodymium, Samarium
PGMs = Platinum Group Metals	Iridium, Palladium, Platinum, Rhodium, Ruthenium

* Carbon is regarded in several forms, for example, BGS included graphite and diamond, EU Commission included coking coal and natural graphite, OECD mentioned natural graphite, and AG, CG, GI, and US DoI chose graphite.
[1] baryte instead of barium.

The Chinese government released National Mineral Resources Planning in 2016 and classified 24 minerals as critical and strategic within three categories (energy, metal, and non-metal). However, this government report was only available in Chinese; thus, the alternative publication in English was used to obtain the list of critical minerals from the Chinese government [41].

The European Commission has published a report on critical raw materials every three years since 2011, and the latest report in 2023 has been mostly referenced. Based on market fluctuations, progress of technology, and production status, the list changes every three years in the reports. The critical raw metals (CRMs) in Table 2 are notably vital in Europe because of their high demand in various industries and the advancement of green and high technology. The two main parameters used by the EU in the methodology to assess the criticality of metals since 2011 are economic importance (EI) and supply risk (SR), which comprise several factors. The EU Commission is trying to ensure the consistency of the assessment throughout the years [39].

The Ministry of Mines of the Government of India published Critical Minerals for India [44]. The report included a global overview, five factors of the value chain of critical minerals, the methodology of classification, and so on. An interesting and meaningful exercise by the Government of India that was included in this report was a comparison of critical minerals chosen by different ministries in India through inter-ministerial consultation. For example, the Department of Science and Technology and the Department of Pharmaceuticals notably chose different minerals as critical; hence, a broad spectrum of critical minerals was reviewed and chosen [44].

The Organisation for Economic Co-operation and Development has 38 countries from around the globe as members, building international standards and policies and solving various issues together [45]. Hence, the OECD's view on critical and strategic materials is significant since the organisation represents a broad perspective regardless of geological factors. In terms of critical metals, the OECD is mentioned in the case study of critical metals and mobile devices for sustainable material management. The organisation acknowledges that the definition of critical metals is subject to change over time due to factors such as geography and technology [46].

In 2020, the Energy Act of 2020 was passed by the United States Congress to boost clean energy and climate policy in the US. Following the Energy Act of 2020, the Secretary of the DoI, who is also the director of the US Geological Survey (USGS), published the draft list of critical minerals in the Federal Register. This 2021 list was consulted by the Department of Agriculture, Commerce, Defence, and Energy, and the United States Trade Representative [42]. In 2022, the Final List of Critical Minerals was published [47]. The recent reports on the critical materials from the Department of Defence (DoD) and Energy (DoE) referred to the list from the DoI [48,49]. Even if a commodity has no supply risk or trade exposure and then a single producer in the US becomes unable to produce the commodity, there is a chance to be exposed to the supply risk, which was categorised as Single Point of Failure (SPOF). With the SPOF method, Be, Ni, and Zr are included in the draft list of critical minerals. In cases where there was insufficient data for a quantitative assessment of a certain commodity, a qualitative evaluation of supply and demand was conducted. As a result, caesium (Cs), rubidium (Rb), Sc, and the REEs remained on the list [50].

5. Recovery of Critical and Strategic Metals from Non-Ferrous Slags

As demand for critical and strategic metals keeps increasing from various industries, securing primary and secondary sources of critical and strategic materials and having feasible extraction technologies are crucial for nations' economies and defence. Hence, in the following chapter of this paper, we would like to introduce the potential of extracting critical and strategic metals from the by-product of the smelting process, non-ferrous slags. Indeed, because the recovery of critical and strategic metals from slag materials is still in its infancy, the available literature is limited. Thus, tin, copper, nickel, vanadium, and titanium slags were studied. Also, it is worth noting that the emphasis on base metals will be less than that on non-base metals since the production and supply of base metals from the primary source are more stable than the other critical and strategic metals.

5.1. Tin Slag

Tin slag is a by-product of the tin smelting process. Tin slag has drawn broad attention since it contains a significant amount of critical and strategic metals, representatively Nb and Ta [51]. It can be considered a secondary source of Nb and Ta, as there are a limited number of primary ores of those metals [52]. Several studies [52–57] have attempted to recover valuable metals from tin slag using pyro- and hydrometallurgical methods. The chemical composition of tin slag, including the critical and strategic elements, is shown in Table 3. Tin slag became a feasible candidate for secondary sources of Ta when the market price increased [58], and it can happen any time soon for other valuable metals or commodities as well.

As shown in Table 3, tin slags contain not only Sn, Nb, and Ta but also critical and strategic metals such as Ti, W, V, Cr, Ni, and REEs. The figures in Table 3 are mostly average concentrations of samples to avoid including all the samples from the same countries and regions in each paper [52,59]. The concentrations of Nb and Tb vary significantly depending on the origin of the sample: Nb 70.89 mg/kg to 14.1% and Ta 1.75 mg/kg to 20.8% as oxides. Thus, many sources actually have a high grade of Nb and Ta: Nb + Ta > 8% high, 5% < Nb + Ta < 8% low, and Nb + Ta < 5% extremely low [58]. Also, the concentrations of tin slags from Smelterskop and Elandsberg, South Africa are ancient samples from 1650 to 1850 CE. The geologists estimated that 180,000 tonnes of tin were mined in these regions before 1905 [59]. Thus, considering the grade of Sn in these regions is 11.10% SnO to 25.28 SnO_2, it might be feasible to extract Sn from the historical tin slags in South Africa.

Table 3. Compositions (%) of critical and strategic elements in tin slags.

Origin (Region)	Nb_2O_5 [Nb]	Ta_2O_5 [Ta]	$(Ta, Nb)_2O_5$	TiO_2 [Ti]	SnO_2 [Sn]	MgO [Mg]	Al_2O_3 [Al]	MnO [Mn]	WO_3 [W]	V_2O_5 [V]	Cr_2O_3 [Cr]	Ni_2O_3 [Ni]	Ce + La	References
N/A	9.35	2.6	11.95	5.9	-	5.3	6.56	-	-	-	-	-	-	[52]
N/A	5.2	7.5	12.7	1.3	0.7	-	11.2	3.7	-	-	-	-	-	[53]
Nigeria	14.1	20.8	34.9	-	0.37	1.2	5.1	1.7	-	-	-	-	-	[55]
Malaysia	3.43	3.05	6.48	7.97	1.33	-	10.42	0.7	1.4	-		-	-	[56,60]
	4	4	8	11	0.5	0.5	9	0.5	8	0.5	-	-	-	[61]
Indonesia	-	-	2.7	13.3	0.8	-	8.0	0.4	0.0	-	-	-	-	[57,60]
Indonesia	0.33	0.64	0.97	11.92	-	-	11.7	-	-	-	-	-	-	[62,63]
(Bangka)	-	-	-	2.38	1.78	-	0.52	-	-	-	-	-	1.7	[63]
Spain	-	-	18.7	7.3	0.1	-	6.9	1.7	0.1	-	-	-	-	[57,60]
South Korea	-	-	10.8	17.8	0.8	-	4.7	0.7	0.1	-	-	-	-	[57,60]
Thailand	-	-	24.9	15.5	0.7	-	7.0	2.1	1.0	0.14	0.65		-	[57,60]
DR Congo	-	-	12.7	1.3	0.7	-	11.2	3.7	0.2	-	-	-	-	[57,60]
England	-		-		16.44	2.21	10.60	0.68	7.31	-	-	-	-	[64]
Zimbabwe	[29.9]	[27.3]	-	[0.16]	-	-	[0.17]	[2.95]	[0.41]	[0.06]	[0.03]	[0.13]	-	[65]
Australia	5.5	10	15.5	9	-	-	-	-	-	-		-		[62]
South Africa	7	5	12	2	-	-	-	-	-	-	-	-	-	[62]
(Smelterskop)	70.89 *	1.75 *	72.64 *	1.4	25.28	3.09	11.86	0.11	0.34	0.11	0.045	20.41 *	456.9 *	A, [59]
(Elandsberg)	0.1	-	0.1	8.61	11.19 [a]	2.37	7.34	0.19	-	0.13	0.13	0.02 [b]	-	A, [59]
Brazil	-	-	-	0.8	-	6.2	11.8	0.7	-	-	-	-	-	[66]
Brazil	[0.3]	[0.5]	-	[1.0]	[2.0]	[4.0]	[2.0]	[0.6]	-	-	-	-	-	[51]

* is mg/kg, not %. A is the ancient sample. [a] and [b] represent different forms of oxides in this case, [a] is SnO and [b] is NiO.

Nb and Ta are the main products that can be extracted from tin slags. As shown in Table 4, several studies [52–54] showed that pyrometallurgical methods, such as chlorination and carbochlorination, were mostly adapted for the recovery of Nb and Ta. For the hydrometallurgical methods, successive acid and alkali leaching were studied [55–57,67]. Conventional methods of extracting Nb and Ta from tin slags are hydrofluoric acid leaching or successive leaching of hydrofluoric and sulphuric acids, followed by carbochlorination. So, the most suitable process in the past would have been a combination of pyro- and hydrometallurgical methods. However, they are less utilised due to economic and environmental issues these days [53]. In June 2019, the latest study by [57], showing the best outcomes of grade and recovery of Nb and Ta, developed the acid–alkali–acid leaching process presented in Figure 1. Although a greater recovery of Nb and Ta is desirable, the grade of the product is outstanding compared to pyrometallurgical methods such as carbochlorination and roasting.

Table 4. Extraction methods of critical and strategic metals from tin slags (Condition and recovery given for bolded method).

Samples	Methods (Pyrometallurgical, Hydrometallurgical and Combined)	Optimum Conditions	Recovery (Grade)	References
Tin slag	- (Pre-treatment) Leaching - Chlorination - **Carbochlorination**	LGC (Low grade concentrate) 1000 °C, 24 h	Nb_2O_5 98.4% Ta_2O_5 99.4%	[53]
Tin slag	- Hydrochloric acid leaching - **Carbochlorination** - Carbon-tetrachloride chlorination	20% Charcoal 900 °C, 40 min	Nb_2O_5 79.8% Ta_2O_5 64.5% TiO_2 74.3% Converted	[52]
Tin slag (Thailand, Zaire, Indonesia)	- Successive acid and base leaching (HF, HCl and NaOH) - **Carbochlorination**/halogenation (Cl_2 + N_2 + CO)	Leaching NaOH 2 N HCl 2 N HF 1.4 N, 70 °C, 1 h Solid/Liquid: 50 g/L Carbochlorination HGC, 500 °C, 24 h	Nb_2O_5 99.71% Ta_2O_5 98.36%	[54]
Zimbabwe tin slag	**In situ carbothermic reduction**	Reducing agents: Collie coal, 1275 °C, 30 min	NbC 31.81% Nb_2O_5 48.11% TaC 12.49% Ta_2O_5 10.37%	[65]
Tin slag (Congo, Indonesia, Malaysia, South Korea, Spain, Thailand)	**Acid-Base-Acid leaching**	HCl 2 M HF 1.2 M NaOH 2 M 70 °C, 4 h, Solid/Liquid: 50 g/L	$(Ta + Nb)_2O_5$ 85.6% (63.3%)	[57]
Malaysian tin slag	- Alkali pre-treatment (Caustic leaching, alkali pugging, caustic fusion) - Hydrochloric acid leaching	Alkali processing Slag/NaOH: 1/2, 250 °C, 4 h Acid Leaching HCl: 3.25 N, 50 °C, 15 min	Nb_2O_5 86.5% (10.6%) Ta_2O_5 95.5% (10.4%)	[56]
Indonesian tin slag	Roasting Water-quenching Leaching (**HF**/HCl and NaOH)	Roasting 900 °C, 2 h Leaching HF 8%, 2 h HCl 6 M, 2 h NaOH 10 M, 20 h	Nb_2O_5 91% (1.34%) Ta_2O_5 87.5% (0.66%)	[67]
	Roasting Water-quenching Successive leaching (**NaOH** and $HClO_4$)	−200 + 250 mesh Solid/Liquid: 1/10 Roasting 900 °C, 2 h Leaching NaOH 8 M, 2 h, 25 °C	Ce (4.35%) La (1.45%) TiO_2 (3.23%) Sn (6.49%)	[63]
	Roasting Water-quenching **Alkali and acid leaching**	Roasting 800 °C, 1 h −100 + 150 mesh Leaching Solid/Liquid: 1/10 NaOH 5 M, 20 min HCl 5 M, 50 min 25 °C	Nb_2O_5 (1.11%) Ta_2O_5 (1.56%) TiO_2 (21%)	[62]

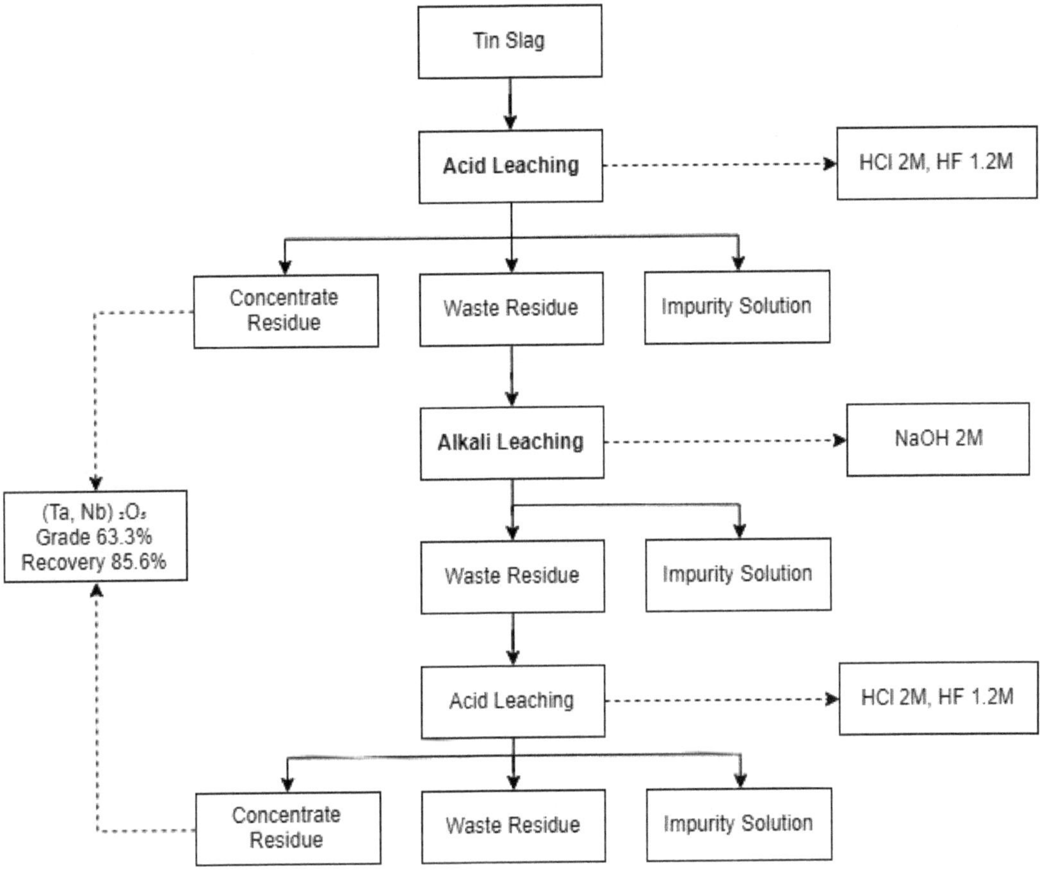

Figure 1. The optimum process for tin slag to recover niobium and tantalum adopted from [57].

Tin slags also contain other critical and strategic metals, such as Ti and REEs, in addition to Sn. Since current studies mostly focus on recovering Nb and Ta from tin slags, a knowledge gap exists in the recovery of other critical and strategic metals. Hence, exploring a suitable process to extract other critical and strategic metals from tin slags is a beneficial and desirable provision for the mineral industry. Other research gaps and current drawbacks of utilising tin slags include the following: (1) low grade and recovery of critical and strategic metals; (2) separation of valuable metals that have chemical similarities; and (3) silica gelation. Since the latest study [57] was able to produce a commercial grade product (63.3%), the research is progressing with technology. It will be favourable to examine green reagents, for instance, organic acids.

5.2. Copper Slag

Copper slag is a by-product of copper production. As much as copper mines have been active around the world, an enormous amount of copper slag has been accumulated in dumps. Copper slag is known as a suitable material for construction, geotechnical applications, road pavements, and so on due to its physical and chemical properties [68,69]. However, this review attempts to highlight the potential economic benefits of copper slag as a great secondary source of valuable metals since copper slag contains a significant amount of critical and strategic metals, as shown in Table 5 [70].

Table 5. Compositions (%) of critical and strategic elements in copper slags (F: Furnace slag, C: Converter slag, A: Ancient slag).

Origin (Region)	Co	Ni	Al_2O_3 (Al)	MgO (Mg)	MnO (Mn)	TiO_2 (Ti)	V	Cr	Mo	Sr	Y	Sb	Sn	References
Germany	0.013	0.012	(7.27)	(3)	(0.26)	(0.42)	0.16	0.026	0.021	0.03	33 ppm	-	-	F, [71]
Turkey	0.023	-	-	-	-	-	-	-	-	-	-	0.021	-	[72,73]
	0.51	35 ppm	-	-	-	-	-	-	-	-	-	-	-	A, [74,75]
	0.095	0.065	-	-	-	-	-	-	-	-	-	-	-	F, C, [76]
	0.104	0.015	2.4	2.82	(0.042)	-	-	-	-	-	-	-	-	F, [77]
	0.36	0.045	-	-	-	-	-	-	-	-	-	-	-	C, [78]
	0.43	0.035	-	-	-	-	-	-	-	-	-	-	-	[79]
Poland	0.145	-	5.73	4.99	-	-	0.123	-	0.077	-	0.026 (ΣREE)	-	-	F, [80,81]
	0.063	-	12.45	6.48	-	-	0.012	-	0.0296	-	0.031 (ΣREE)	-	-	F, [80,81]
	-	-	14.7	2.42	0.35	0.69	-	-	-	-	-	-	-	A, [81,82]
	5.8	0.4	-	-	-	-	-	-	-	-	-	-	-	C, [83]
China	0.16	0.53	(3.90)	(3.53)	-	-	-	-	-	-	-	-	-	[5]
	0.49	0.27	1.36	1.34	-	-	-	-	-	-	-	-	-	C, [84]
	0.60	0.23	-	-	-	-	-	-	-	-	-	-	-	C, [85]
	4.09	0.043	(3.22)	(2.15)	(0.14)	(0.19)	-	0.075	0.01	0.012	-	-	0.14	[86]
Congo	0.72	-	(2.56)	(2.53)	-	-	-	-	-	-	-	-	-	[87]
	1.7	-	6.0	5.6	-	-	-	-	-	-	-	-	-	F, [88]
	0.789	79 ppm	8.12	3.36	0.074	0.53	-	0.185	0.017	-	-	8.61 ppm	19 ppm	[89]
Namibia	0.051	73 ppm	5.22	3.09	0.2	0.22	-	-	-	-	-	0.073	-	[90]
South Africa	0.032	0.024	6.61	4.52	0.13	0.78	0.03	0.11	68 ppm	0.023	0.04 (ΣREE)	-	15 ppm	[91]
Finland	0.04	0.05	3.8	1.0	(0.07)	-	-	-	0.07	-	-	0.04	0.1	F, C, [92]
Chile	-	-	(0.31)	-	-	-	-	-	0.3	-	-	-	-	[93]
	-	-	25.75	1.15										F, [94]
	-	-	2.72	0.83	-	-	-	-	-	-	-	-	-	F, C, [95]
India	0.48	1.98	(0.08)	(2.65)	-	-	-	-	-	-	-	-	-	C, [96,97]
	0.19	0.23	(1.58)	(0.97)	-	-	-	-	-	-	-	-	-	F, [98]
Italy	-	-	4.66	0.89	1.57	0.19	-	-	-	-	-	-	-	A, [99]
N/A	0.1	0.027	-	1.035	-	-	-	0.024	-	-	-	0.043	-	F, C, [100]
	0.44	<0.1	(0.83)	(0.38)	-	(10 ppm)	-	0.16	<0.1	-	-	50 ppm	<0.5	C, [101]

There are two forms of copper slag, which vary based on their cooling processes. Air-cooled slag and water-quenched slag show different properties. Water-quenched slag has a more granular form, like sand. Thus, it can be suitable for use in the construction industry. Air-cooled slag is slow-cooled and can be suitable for extracting valuable metals since it may contain various forms of minerals due to crystallisation with enough time [6,68,69]. However, slow-cooled slags tend to have more dense structures than fast-cooled slags, so it is sometimes harder to liberate valuable metals from slow-cooled slags.

Every ton of copper production generates 2.2 tonnes of copper slag [8]. Total reserves of copper in the world are estimated at 1 billion tonnes in 2024, and the global mine production of copper in 2023 was 22 million tonnes [102], which would have generated 44 million tonnes of slag. The largest production occurs in Chile, which was 5 million tonnes in 2023. Chile is followed by Peru, Congo, China, the USA, Russia, Indonesia, Australia, Zambia, Mexico, Kazakhstan, Canada, Poland, and others in terms of copper mine production [102]. As is shown in the active production of copper, a significant amount of copper slag keeps being dumped worldwide. It would be beneficial to recover valuable metals from copper slag for those countries that have significant amounts of dumped slag and will constantly generate slag in the future.

The amount of copper slag that is dumped in known and unknown places would be immeasurable. For example, around 2.5 million tonnes of ancient Turkish copper slag from the 13th century have been stored in the northern part of Turkey. The ancient slag also was adopted to develop the process by researchers. Based on the chemical analysis, this slag has 0.5% of cobalt, which can produce around 12,500 tonnes of cobalt [74,75], which is equivalent to 337 million US dollars (US $27,000 per tonne in June 2024) [103]. There is a literature [104] that reports a pilot-test work performed on a copper reverberatory furnace slag in Zambia and the production of cobalt from the copper slag. Around 20 million tonnes of dumped slag that contained 0.66% cobalt were processed through the furnace, which commenced in 2001.

Methods to recover critical and strategic metals from copper slags are presented in Table 6. The main valuable metals are Cu, Co, and Ni. The extraction of base metals from copper slag was excluded since base metals' criticality is lower than that of critical and strategic metals [43,105]. To recover Co from copper slags, roasting and leaching were adopted, along with sulphurisation, flotation, precipitation, and so on. More details on the generation, characteristics, utilisation, environmental impacts, and recycling technology of copper slag, regardless of the critical and strategic elements, can be referred to in [8,9,12].

Several attempts have been made to recover Co from copper slag, mainly by hydrometallurgy approaches. They are showing remarkable results, considering that traditional methods mostly include roasting or other pyrometallurgical methods. After leaching, researchers also adapted separation and purification methods, such as solvent extraction and precipitation. Figure 2 is the process flowsheet to recover Co selectively using leaching with sulphuric acid and sodium chlorate as an oxidant and precipitation with calcium hydroxide, and it produced 98% recovery of Co, which is an outstanding result considering it was conducted under atmospheric pressure [86].

Another feasible process is shown in Figure 3. Banza and colleagues investigated a combined process of hydrometallurgical methods to extract Co, Cu, and Zn. Leaching was conducted in a sulphuric acid medium with hydrogen peroxide as an oxidant. Then, Fe was separated through precipitation, and Cu was extracted by solvent extraction. Finally, Co and Zn were recovered from the solution by solvent extraction with D2EPHA, and they were stripped separately. The overall recovery of Co was 90%, which indicates it is promising to take into consideration for adapting hydrometallurgical methods to upgrade the copper slag for the extraction of critical and strategic metals [87].

Table 6. Extraction method for critical and strategic metals from copper slags.

Samples	Methods	Optimum Conditions	Recovery (Grade)	References
Turkish Copper Slag	Sulphurisation Roasting Water leaching	−100 mesh (150 µm) Sulphurisation: Closed system, gas mixture; 6 g H_2S + 25 g H_2O, 130 °C, 1 h, solid-gas interaction, Roasting: 600 °C, 6 h, Water leaching	Co 98.8%	[106]
	Roasting Water leaching	−100 mesh (150 µm), Roasting: 200 °C, Sulphating agent; 1.5 stoichiometric amount H_2SO_4, 1 h, muffle furnace, Water leaching	Co 96%	[107]
	Pre-roasting Roasting Water leaching	−63 µm pyrite, various for slag Pre-roasting: 550 °C, 1 h. Roasting: 550 °C, 3 h, pyrite/slag 0.25 Water leaching: 15 min, ambient temperature, pulp density 10%	Co 98.4% Ni 22.2%	[108]
	Roasting Leaching (Microwave acid digestion)	−150 µm, Roasting: Slag: Pyrite: Copper concentrate 3:6:6 (220 g), iron powder 2.2 g, 600 °C, 5 h, closed medium Leaching (Microwave acid digestion): 0.2 g dried sample, 1450 W, 37% HCl 2.5 mL + 65% HNO_3 7.5 mL	Co 98.4%	[73]
	Acid roasting Leaching	−0.1 mm, Acid roasting: H_2SO_4, 2 h, 150 °C, acid/slag 3:1 Leaching: Hot water, 70 °C, 1 h	Co 87%	[76]
	Flotation Leaching	−150 + 300 mesh Flotation: Collector; X−231 100 g/t, Na_2S 300 g/t Leaching: HCl 150 g/t, 2 h, 80 °C, S/L 1/10	Flotation tail Sn 73% (2.58%)	[109]
Turkish ancient · copper slag	Flotation Roasting Leaching	−0.1 mm, Flotation: Na_2S 300 g/t, Aero 211 200 g/t, pH 7~7.2, Pulp density 15%, 1300 rpm Roasting: 500 °C, 1 h, pyrite: slag 3:1, Leaching: H_2SO_4 10 g/L	Co 86.5%	[74]
	Leaching	H_2SO_4 120 g/L, 2 h, 60 °C	Co 90%	[75]
Indian copper slag	Roasting Leaching Precipitation	−75 µm, Roasting (1st and 2nd): 1 stoichiometric amount of H_2SO_4, 150 °C, 1 h/650 °C, Leaching with water Precipitation (iron removal): NH_3 or CaO, pH 2.5–3	Co 95% Ni 95%	[96]
	Pressure leaching	Pulp density 10%, H_2SO_4 0.35 N, 130 °C, 0.59 MPa, 4 h, 840 rpm	Co >95% Ni >95%	[97]
	Reductive roasting Leaching	Reductive roasting: 10% furnace oil, 850 °C, 2 h Leaching: 1.25 stoichiometric amount of $FeCl_3$, 2 h	Co 80% Ni 95%	[98,110]
Chinese copper slag	Reductive—sulfidising reduction Magnetic separation Leaching	Reductive—sulfidising reduction: Product—Cu-Co-Fe matte (Start material), Magnetic separation: 40 mT, Feed rate 100 mL/min Leaching: H_2SO_4 1.15 stoichiometric, 1 h, 80 °C, 500 rpm, 10% (*w*/*v*)	Co 95.57%	[111]
	Reduction-roasting	Reduction-roasting: $MoSi_2$ electric furnace (12 kW), 1350 °C, 1 h, Reducing agent; coke powder 7%, Curing agent; pyrite 30%	Co 95%	[84]
	Reduction smelting	Reduction smelting: 1350 °C, 2.5 h, reducing agent; activated carbon 5%, Modifiers; TiO_2 5%, CaO 4%	Co 94.02% (1.76%)	[85]
	Leaching Precipitation	−180 µm, Leaching (20g slag): H_2SO_4 17 g, oxidant; $NaClO_3$ 2.5 g, 95 °C, 3 h Neutralisation (Precipitation): Ca $(OH)_2$ 4.3 g, pH 2, 1 h	Co 98%	[86]

Table 6. *Cont.*

Samples	Methods	Optimum Conditions	Recovery (Grade)	References
German Copper slag	Sorting Roasting Leaching Precipitation Complexation Solvent extraction	−100 μm Sorting: Crystalline: Vitreous 3:1 Leaching: 37% HCl, 70 °C, 1 h, pH 2.5 (start), Solid/liquid 100 g/L Precipitation: NH_4HCO_3, pH 7 Precipitate complexation: $NH_4OH + H_2O_2$, pH 10 Solvent extraction: H_2SO_4	Various products	[71]
Congolese copper slag	Oxidising leaching Solvent extraction Precipitation Solvent extraction	−100 μm Oxidising leaching: 500 kg/t H_2SO_4 + 62.5 L/t H_2O_2, 70 °C, pH 2.5, 2 h, S/L 10% Solvent extraction: Copper—12% *v/v* LIX 984 in Kerosene, pH 2.5, A/O 1:1 Precipitation: Iron—300 g/L $CaCO_3$, pH 3.5, 85 °C Solvent extraction: Zinc and Cobalt—20% *v/v* D2EPHA in kerosene, 25 °C, pH 3.5, A/O 1:1 Cobalt stripping—H_2SO_4, 25 °C, pH 2.5, Zinc stripping—H_2SO_4, 25 °C, pH 1	Co 90%	[87]
Polish copper slag	Reduction roasting Electrolytic dissolution Ammoniacal leaching Electrowinning	Reduction roasting: electric furnace, 80–90 V, 900–950 A, fluxing agents; limestone 8.6% + dolomite 8.6%, reducing agents; coke breeze 2.7% + pig iron 2.0%, 1425–1570 °C Electrolytic dissolution: ammonia-ammonium chloride electrolyte, $NH_3 + NH_4^+$ 2.5 M, Cl^- 1 M, pH 10.1, circulation rate 15 dm^3/h, 50 °C, anodic current density 350 A/m^2 Ammoniacal leaching (for anode slime from previous stage): $NH_3 + NH_4Cl$ Electrowinning: copper—pH 0, cobalt—pH 6	Co (92%)	[83]
Chilean copper slag	Reduction (Slag cleaning) Water quenching Drying Milling Reduction (Removal of carbon contamination)	Reduction (Slag cleaning): 800 g power mixture; SiO_2, CaO, Al_2O_3 and MgO, Induction furnace (10 kHz, 15 kW), Reductant; graphite, Pure argon atmosphere, Modifier; TiO_2 11 wt% Water quenching Drying: Muffle furnace, 100 °C Milling: −75 μm Reduction (Removal of carbon contamination): Muffle furnace, 8 h, 700 °C	Recovery Co/Recovery Fe ratio 1.5	[112]
Copper converter slag	Curing Leaching Cementation Precipitation	−75 μm, Curing: H_2SO_4, 2 h, 85 °C Leaching: 1.2 times of stoichiometric amount H_2SO_4, 90 min, 75 °C Cementation (Cobalt and Zinc): Cementation reagent; Manganese, Mn/Co molar ratio 7, initial pH 3, 45 min Precipitation (Cobalt sulphide): initial pH 4, 90 min, 100 °C, Precipitate reagent; Barium sulphide, BaS/Co molar ratio 3	Co 98%	[101]

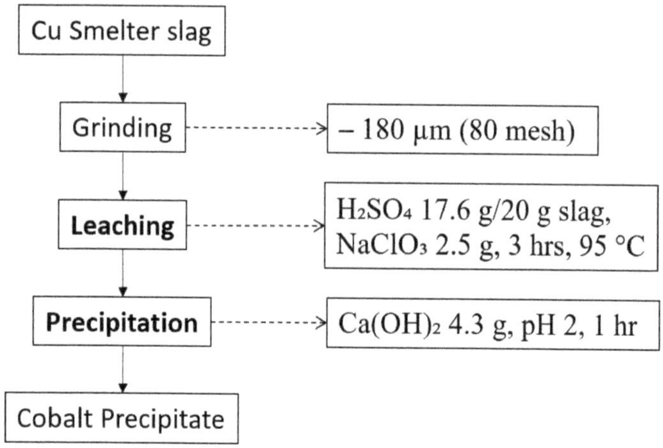

Figure 2. Process flowsheet of copper slag adopted from [86].

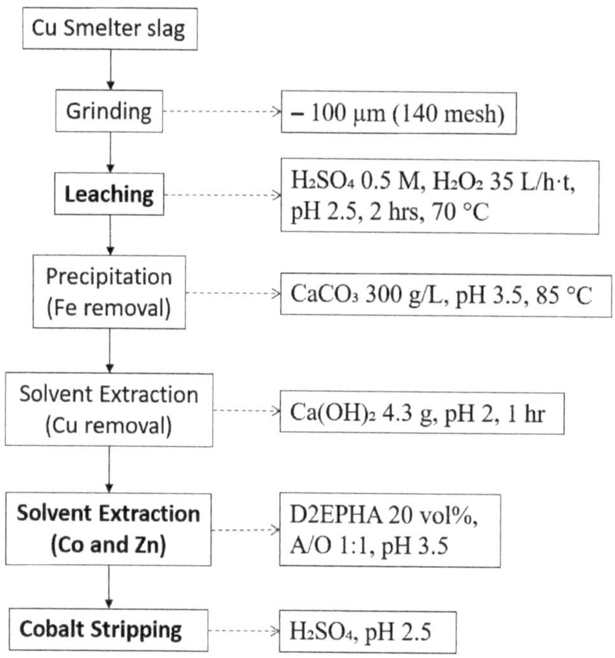

Figure 3. Process flowsheet of copper slag adopted from [87].

A study was also conducted to recover Sn from copper slag. As shown in Table 2, Sn is one of the critical and strategic metals in copper slag. The process includes size reduction and classification, flotation, and leaching. In flotation, Cu was recovered as concentrate and Sn remained in tailing. In the leaching test, the recovery of Sn was not as efficient as that of Cu since the process was focused on Cu extraction. Consequently, there is still a research gap in the recovery of Sn from copper slag [109]. Including the processes that were mentioned above, hydrometallurgical methods are listed in Table 6 to recover critical and strategic metals from copper slag [97,98,101,113].

As one of the hydrometallurgical methods for extracting critical and strategic elements from copper slag, bioleaching has been studied for the past decades. Bioleaching is considered an environmentally friendly and sustainable process and is a gradually evolving technology [114]. In bioleaching, microorganisms play a key role as extractants [115]. A recent study by Mikoda et al. (2019) studied the feasibility of bioleaching assisted by *Acidithiobacillus thiooxidans*. The optimal conditions for this study were a particle size of 0.25–0.5 mm and a pulp density of 1% for 28 days. With these conditions, high recoveries of Co, Mo, REEs, and V were achieved: shaft furnace slag—Co 100%, Mo 44%, REE 70%, and V 70%; and granulated slag—Co 95%, Mo 70%, REE 99%, and V 93% [80]. Another study from the same research group suspected that the bioleaching of historical copper slag could create up to $135 worth of metals from a ton of slag. Hence, extracting values from historical copper slag can be an economically and environmentally positive option [116]. A broader perspective has been implemented by Lee and Pandey (2012), who reviewed the bioprocessing of solid wastes, including copper slag, for metal extraction [10]. Using a mixed culture that contains *Acidithiobacillus* and *Leptospirillum* with 10 g/L sulphur (S) and 4.4 g/L ferrous sulphate (Fe^{2+}) at a pulp density of 10 g/L and a pH of 1.5, in 42 days, 100% recovery of Ni was achieved [117]. Mehta and Pandey [118] recovered 64% of cobalt and 50% of nickel from copper converter slag in 80 days with the optimum conditions of pH 2, pulp density 1/20, and particle size −75 μm using *Thiobacillus ferrooxidans*. Despite significant research efforts aimed at developing bioleaching processes, most of the experiments have been limited to the laboratory scale [10]. Since the new extraction processes should minimise the impacts on the environment and pursue economic profit, bioleaching is considered a green technology for the mining industry.

Pyrometallurgical methods tend to be used frequently owing to effective sulphurisation or reduction. Thus, there have been various implementations of pyrometallurgical methods on copper slag. A recent study by Yuksel and Teğin [73] reported a huge improvement in pyrometallurgy by reducing the temperature for roasting, which saves energy as well as the processing budget. A total of 98.4% of Co from copper slag was recovered by roasting at 600 °C for 5 h with a 3:6:6 ratio of slag: pyrite: copper concentrate and iron powder [73], whereas other studies [84,85] required 1350 °C to extract around 95% Co. Depending on the methods, Co remains in the depleted slag after roasting [84], or a cobalt-bearing alloy can be produced after reduction [85]. In these cases, recoveries of Co were 95% and 94.02%, respectively. There is also a study to identify the most suitable modifier to selectively recover Co from copper slag when the reductant is graphite. It turned out to be more titanium dioxide than calcium oxide or calcium fluoride [112].

One of the greatest achievements of exploiting slag as a secondary source is the pilot test work that was later conducted at a plant in Zambia in 2001. A 40 MW DC arc furnace was built to extract cobalt from the 20 Mt reverberatory furnace slag dump. The feed was specifically blended with slag, lime, and coal, and the tapping temperature on average was 1500 °C [104]. This example indicates that it is economical and feasible to extract valuable metals from slags, along with providing a chance to clean slag if required.

Various flowsheets combining the pyrometallurgical and hydrometallurgical methods have been widely explored to extract critical and strategic metals from copper slag. A few cases of studies in pyrometallurgy required water leaching after roasting, so they were categorised as combined methods [5,106–108]. A novel process was developed by Yin, Xing [111] that includes smelting, magnetic separation, and leaching with sulphuric acid. Through reductive-sulphurising smelting, a Cu–Co–Fe matte was produced and then processed under a magnetic separator. Magnetic concentrate possessed cobalt-bearing phases, which include Co–Fe alloy, and non-magnetic substances had copper-bearing phases. Subsequently, sulphuric acid leaching at 80 °C for 1 h was conducted for magnetic substances. The overall recovery of Co was 95.57% after this series of processes [111]. A research group in India investigated various methods to extract Co and Ni from copper slag: roasting with acid, water leaching and precipitation [96], pressure leaching [97], reductive-roasting, and leaching with ferric chloride [110]. The conditions of these methods are listed in Table 6. These methods showed remarkable results, with recoveries exceeding 95% of Co and Ni. Several attempts have also been made to recover Mo from copper slag by a sequence of selective reduction, roasting, and leaching with sulphuric acid [93] and by oxidative roasting followed by sulphuric acid leaching [119]. The former study achieved enrichment of Mo by 8.9 (final/initial grade: 2.67/0.3), and the latter recovered 80% of Mo. One of the most comprehensive flowsheets for processing copper slag is presented in Figure 4. It encompassed from reduction roasting to electrowinning and produced a 92% purity of Co [83]. It indicates that a combined flowsheet of hydro- and pyrometallurgical methods for extracting critical and strategic metals such as Co can produce high-purity concentrates. However, the feasibility, profitability, and viability of the process should still be assessed before applying to industry.

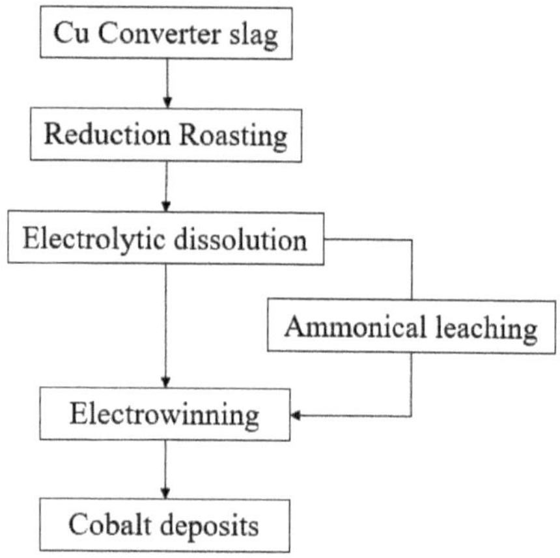

Figure 4. Flowsheet of a selective process for cobalt from copper slag adopted from [83].

5.3. *Nickel Slag*

Nickel slag is a smelting waste from nickel production [120]. Slag is usually stored in dumps since they are considered as waste materials [121]. Furthermore, nickel slag contains critical and strategic metals, for example, Ni and Co, as shown in Table 7. One of the potential problems with dumped slags is that they may contain potentially toxic elements (PTEs). In case of rain, PTEs can be leached out from the slag and exposed to the environment. Therefore, the extraction of valuable metals from nickel slags is not only a great secondary source of critical and strategic metals but is also able to make the slag cleaner for the environment.

Nickel slag can be categorised based on its production stages and cooling methods: (1) furnace or converter slag and (2) slow-cooled or fast-cooled slag. They behave differently in a chemical reaction and have dissimilar crystal structures. In the pyrometallurgical process, the slag from the smelting phase is called furnace or smelter slag, and the other slag from the converting phase is distinguished as converter slag. Usually, converter slag contains a higher concentration of valuable metals than furnace slag since converting deals with a higher grade of product. For example, Ni in converter slag is 2.87–4.80%; however, the primary ores contain 1.05–2.3% of Ni [122]. For this reason, converter slag is normally going back to the circuit for processing.

Due to cooling time and methods, slow-cooled and fast-cooled slags show different crystallisations. Slow-cooled slag contains well-developed crystalline, fayalite, and minor spinel. On the other hand, fast-cooled slags are more homogeneous in terms of metal distribution and contain amorphous iron silicate glass [7].

Nickel slag contains critical and strategic metals such as Ni, Co, Ti, Al, Mg, Mn, V, and Cr, as shown in Table 7. Nickel slag also has base metals, for example, Fe, Cu, Si, etc.;

however, they were excluded from Table 7 because they are regarded as less critical due to their stable supply. In recent years, there has been an increasing interest and attempt to recover Ni and Co from nickel slag in the industry. Hence, this review mostly focuses on the extraction of Ni and Co from nickel slag based on the publications.

Table 7. Compositions (%) of critical and strategic elements in nickel slags (SC: Slow-cooled slag, FC: Fast-cooled slag, C: Converter slag, F: Furnace slag).

Origin (Region)	Ni	Co	Ti	Al	Mg	Mn	V	Cr	References
Brazil	0.08	73 ppm	0.04	1.99	17.2	0.29	85 ppm	0.76	[120]
	0.38	0.02	-	0.68	5.76	-	-	1.11	[123]
	1.05	0.243	-	1.66	0.8	0.05	-	-	SC, [122,124]
China	1.48	0.6	-	-	-	-	-	-	C, [125]
	0.9	0.12	-	1.4	5.83	-	-	-	F, [126]
Canada	0.28	0.13	0.14	3.63	1.87	0.04	-	-	F, [127]
	0.66	0.21	0.16	2.62	1.54	0.04	-	-	SC, [7]
	0.31	0.14	0.1	3.73	2.0	0.03	-	-	FC, [7]
	0.19	0.11	0.21	3.97	2.1	0.05	-	-	SC, [7]
	0.05	0.1	0.02	0.78	0.49	0.02	-	-	FC, [7]
Poland	0.55	0.05	0.64 (TiO_2)	11.47 (Al_2O_3)	2.76 (MgO)	0.26 (MnO)	-	-	[128]
	0.02	77 ppm	0.53 (TiO_2)	16.4 (Al_2O_3)	5.52 (MgO)	1.16 (MnO)	-	-	[128]
	82 ppm	5 ppm	0.09 (TiO_2)	2.79 (Al_2O_3)	10.06 (MgO)	0.22 (MnO)	-	-	[128]
N/A	3.64	1.05	-	1.63	0.29	-	-	-	C, [129]
	0.27	0.11	-	3.84	3.81	-	-	-	F, [129]
	0.32	0.16	0.12	2.86	2.57	0.05	0.01	0.05	[121]
	0.16	0.05	0.13	3.64	1.22	0.03	0.01	0.05	[121]
	0.12	0.02	-	6.92 (Al_2O_3)	8.49 (MgO)	0.34 (MnO)	-	-	F, [130]

The concentrations of Ni and Co in nickel slag are from 82 ppm to 3.64% and from 5 ppm to 1.05%, respectively. The chemical composition of critical and strategic metals in nickel slag from various references is shown in Table 7. The maximum Ni grade is 3.64%, from converter slag, which is higher than the primary ore's grade. The highest grade of Co in nickel slag is 1.05%, from converter slag [129]. Nickel slag also contains a reasonable amount of Ti and Cr which implies that nickel slag can be a secondary source of these other valuable metals.

There have been several studies in the literature on nickel slag reporting leaching behaviour [120,121,127,129], selective recovery of valuable metals [122,125,126], characterisation, and mineralogy [7,124,128], and environmental effects [130]. There has been plenty of research on the characteristics of nickel slag in the past, and recently, research has been focusing on the extraction of valuable metals from nickel slag as well. However, there is a limited number of publications on nickel slag.

Ettler, Kvapil [120] investigated the leaching behaviour of metals in nickel slag. The optimum pH was 3 in the range between pH 3 and 12. In this range, chromium shows a U-shaped leaching pattern. Nickel, cobalt, and magnesium have a decreasing trend in concentration from pH 3 to 12, with a low concentration at around pH 12. However, this study focused on the exposure of potentially toxic elements (PTEs) to the environment rather than an extraction method of valuable metals from nickel slag [120]. It is seldom appropriate to apply the result for the recovery of valuable metals to this study since the leaching test was to evaluate the environmental implications when these metals were dissolved. Thus, further studies of leaching behaviour in the extraction process for production purposes are required for a better understanding of the leaching mechanism.

Extraction processes, including the pyrometallurgical methods, were also studied. Huang et al. [125] achieved more than 97% recovery for nickel and cobalt using leaching with 0.3 M sulphuric acid and 600 kPa oxygen partial pressure at 200 °C for 80 min. Pan et al. [126] conducted selective reduction and magnetic separation to recover nickel and copper and demonstrated when the basicity increases, the enrichment of nickel improves as well. As a result, the author recovered 82.2% of nickel with a 3.25% grade.

The optimum method for the recovery of critical and strategic metals from nickel slag is high-pressure and -temperature leaching. There have been attempts to dissolve the target metals at atmospheric pressure, but it was unsuccessful due to several drawbacks, such as slow oxidation of iron, slow filtration, and high acid consumption [121]. Thus, current studies are focusing on high-pressure oxidative leaching, which provides selectivity towards Ni and Co with more than 90% of the recovery.

The common parameters of high-pressure temperature oxidative leaching from various research groups are a choice of acid and temperature. Weak sulphuric acid is a by-product of nickel production; as a result, nickel production companies have a significant amount of acid stored. It is desirable to utilise sulphuric acid to process the slag. One of the reasons that weak acid should be applied is that under atmospheric pressure and temperature, above 0.05 M of sulphuric acid reduces selectivity towards Ni and Co in nickel slag [121]. Thus, employing high-pressure and -temperature leaching with a slightly increased acid concentration is consistent with the literature as shown in Table 8. The temperature of the leaching experiment is mostly 250 °C as shown in Table 8. A high temperature, especially 250 °C, ensures that iron and aluminium do not dissolve to leach the solution at this temperature [7]. With high-pressure and -temperature leaching, an over 90% recovery of Ni and Co could be extracted in several publications which shows a promising result of the hydrometallurgical method.

Table 8. Extraction methods of critical and strategic metals from nickel slags.

Samples	Methods	Optimum Conditions	Recovery (Grade)	References
Chinese slag	High-pressure oxidative acid leaching	0.3 M H_2SO_4, 200 °C, 80 min, 600 kPa, −150 + 74 μm, Solid/liquid 1 g/7 mL	>97% Ni, Co	[125]
Brazilian slag	High-pressure oxidative acid leaching	20% Acid, 250° C, 2 h, 500 kPa, Solid/liquid 25%	>99% Ni, Co	[124]
	High-temperature pressure oxidative acid leaching	20% Acid, 250 °C, 2 h, 200–300 kPa, Solid/liquid 25%, Acid consumption 38.5 kg H_2SO_4/t slag Oxygen consumption 84 kg O_2/t slag	>99% Ni, Co	[122]
Nickel slag	Oxidative pressure acid leaching	H_2SO_4, Acid/slag 0.3, 250 °C, 1 h, 250 kPa	>90% Ni, Co	[129]
Canadian slag	Aqueous sulphur dioxide leaching	Aqueous SO_2 1 M SO_2 (aq.), 200 mL/min, 70 min Leaching 900 mL 1 M SO_2 (aq.) + 9 g slag, 600 rpm, −106 + 75 μm	Co 77% Ni 35%	[127]
Nickel slag	Sulphuric acid leaching	H_2SO_4 0.1 M, 20 h, Solid/liquid 1/10	Co 75% Ni 65%	[121]
Chinese slag	Selective reduction Magnetic separation	Selective Reduction 5% coal, 1200 °C, 20 min Magnetic Separation 131.34 kA/m, 75 μm	Ni 82.2% (3.25%)	[126]
Nickel slag	Magnetic separation	Hand magnet (Low-intensity magnetic separator), −8 + 4.7 mm	Ni 65.92% (1.31%)	[130]

For further development of a feasible process, removing impurities such as Fe, Cu, Si, etc., in other words, improving the grade of Ni and Co, will be one of the major difficulties since target metals in starting materials are low grade. Based on the results of magnetic separation, the combination of high-pressure oxidative leaching and magnetic separation may be an option; however, there is no literature on this combination. Another way to enhance the selectivity of valuable metals can be complexation leaching, which employs a complexing agent in leaching.

Another challenge for the recovery of Ni and Co in nickel slag will be the separation of those metals since they exist together in the leach solution. So far, solvent extraction seems the optimum way to separate Ni and Co using Cyanex as an extractant [131–133]. Since Cyanex 272 is effective in extracting cobalt from nickel, Mubarok and Hanif (2016) [131] used an artificial solution containing only Co and Ni without impurities to separate them. As a result, the authors extracted 99.08% of Co with 10.16% of Ni at pH 5, Cyanex 272 concentration 20% (v/v), and at room temperature. However, from [132,133], it was confirmed again that Cyanex 272 is the optimal extractant and the most commercially used in the industry for the separation of Co and Ni. Further studies on separation focusing on the leach solution from nickel slag are required.

5.4. Vanadium Slag

This chapter discusses the recovery of critical and strategic metals from vanadium slag, including the chemical compositions of vanadium slags, extraction methods, and so on. Notably, most of the literature and sources of vanadium slag are from China. Several studies on the extraction of critical and strategic metals from vanadium slag have increased since the 2010s, which shows the potential of vanadium slag as a secondary source.

Vanadium is widely exploited in ferrous and non-ferrous alloys and is mostly consumed in the iron and steel industry due to its outstanding physical properties, such as fatigue resistance, hardness, and so on [13,134]. Vanadium can be found in vanadium–titanium magnetite, vanadium slag, vanadium stone coal, steel slag, spent catalyst, and fly ashes [135–137]. The main source of vanadium comes from vanadium titano-magnetite as a form of slag. Smelting the vanadium titano-magnetite ore enriches vanadium into solid slag, which is called vanadium slag [138]. Vanadium is mostly associated with ferrovanadium spinel ($FeO{\cdot}V_2O_3$) as a main phase for both vanadium–titanium magnetite and vanadium slag [137]. Vanadium slag contains a significant amount of chromium, titanium, and vanadium. Table 9 shows the chemical compositions of vanadium slags from various origins in China.

Table 9. Compositions (%) of critical and strategic elements in vanadium slags.

Origin (Region)	V_2O_5 (V)	Cr_2O_3 (Cr)	TiO_2 (Ti)	Al_2O_3 (Al)	MnO (Mn)	MgO (Mg)	References
China (Sichuan)	(6.85)	(5.66)	(5.94)	-	(4.81)	(2.10)	[139]
	(8.57)	-	(8.08)	-	(7.05)	(4.06)	[135]
	16.17	1.18	11.97	3.85	8.43	1.95	[140]
	14.3	4.4	7.4	2.1	8.5	3.7	[136]
	8.55	2.29	10.98	-	8.45	3.07	[141]
(Chengde)	13.72	9.19	10.45	1.30	6.73	1.17	[142]
(Liaoning)	(3.3~5.2)	(3.4~5.7)	-	-	-	-	[143]
(Panzhihua)	(7.15)	(2.19)	(6.86)	(1.55)	(5.92)	(1.38)	[144]
	(8.15)	(2.77)	(7.75)	(1.79)	(6.91)	(2.73)	[145]
(Hebei)	10.20	4.15	11.03	2.42	5.22	1.60	[146]
	14.42	9.45	11.19	-	7.03	-	[147]

Table 9. *Cont.*

Origin (Region)	V_2O_5 (V)	Cr_2O_3 (Cr)	TiO_2 (Ti)	Al_2O_3 (Al)	MnO (Mn)	MgO (Mg)	References
China	12.22	5.14	10.82	2.22	5.72	2.95	[148]
	15.44	-	11.45	-	6.67	3.53	[149]
	14.08	-	10.63	1.67	6.50	3.25	[150]
	(12.0)	(6.48)	(0.02)	(0.04)	-	-	[151]
N/A	10.45	3.54	9.75	1.1	5.22	2.15	[152]

Countries that possess vanadium in the world are Australia, Russia, China, South Africa, the USA, and others [13]. In 2023, China mined 68,000 tonnes, while Russia, South Africa, and Brazil produced 35,500 tonnes all together: 20,000 tonnes, 9100 tonnes and 6400 tonnes, respectively. These countries combined have produced approximately 103,500 tonnes of vanadium in 2023, which makes up the world's total production [102,153].

A traditional method to extract V from vanadium slag is sodium salt roasting and water leaching. However, this process can recover only 65 to 85% of the V during leaching due to its poor transfer efficiency [146]. Other well-known methods are calcification roasting, acid leaching, and direct alkaline leaching. Calcification roasting was introduced due to limited access to the sodium slag in Russia, while direct alkaline leaching was suggested to overcome the disadvantages of pyrometallurgical processes [137]. Thus, as can be seen in Table 10, several attempts from predominantly China have been made to overcome the drawbacks and to eventually achieve a higher recovery of V as well as Cr with various methods, such as electrochemical decomposition [146], microwave roasting—leaching [140], and electro-oxidation leaching [135].

Table 10. Extraction methods of critical and strategic metals from vanadium slags.

Samples	Methods	Optimum Conditions	Recovery (Grade)	References
China (Sichuan)	Sulphating roasting Reduction leaching Selective oxidisation Precipitation	Sulphating roasting: H_2SO_4 40 wt%, 160 °C, 2 h Reduction leaching: Na_2SO_3 2%, S/L 1 g/4 mL, 95 °C, 1 h Selective oxidisation: oxidant; CrO_3 6 g/100 mL, $(NH_4)_2SO_4$ 16 g/L, pH 2.0, 95 °C, 2 h - Filtration; 1.5 H, 105 °C - Calcination for vanadium; 550 °C, 2.5 h Precipitation: pH 3.2~3.8 by Na_2CO_3 (Fe removal), pH 8.5 by NaOH at 90 °C (Cr)	V_2O_5 (99.1%) Cr_2O_3 (98.9%)	[138]
	Electro-oxidation leaching	−200 mesh, H_2SO_4 40 wt%, S/L 250 g/L, 75 °C, 4 h, anode current 0.4 A, operating potential 2.8~3.0 V, electrode spacing 20 mm, M (V slag:$MnSO_4$) 2.5:1	V 75.64%	[135]
	Microwave roasting Leaching	−75 μm, Microwave roasting: 2450 MHz, 350 °C, 1 h Leaching: H_2SO_4 250 g/L, 1 h, 90 °C, 350 rpm	V 94%	[140]
	Calcification roasting Acid leaching	−75 μm, Calcification roasting: M (V slag:CaO) 16:1, muffle furnace, 850 °C, 2 h Acid leaching: H_2SO_4 15%, 55 °C, S/L 1:10, 150 rpm	V 93%	[141]
	Direct roasting Soda leaching	−75 μm, Direct roasting: Na_2CO_3 160 g/L, 850 °C, 1 h Soda leaching: 95 °C, 150 min, S/L 1:10	V > 90%	[154]
China (Liaoning)	Salt roasting Water leaching Solvent extraction Precipitation Roasting	Roasting: muffle furnace, sodium salt Water leaching Solvent extraction: LK-N21, mixing 15 min, 20 °C Precipitation: Ammonium salt Roasting: high temperature to produce vanadium pentoxide	V_2O_5 98.5% (99.5%)	[143]

Table 10. *Cont.*

Samples	Methods	Optimum Conditions	Recovery (Grade)	References
China (Panzhihua)	Salt roasting Electro-oxidation leaching	Roasting: sodium salt Electro-oxidation leaching: current density 1000 A/m^2, S/L 1 g/4 mL, 80 °C, 40 min	V 93.67%	[144]
	Salt roasting Water leaching	Roasting: NaCl 30% + Na_2CO_3 20%, 700 °C, 2 h Leaching: 95 °C, 3 h	V 96% Cr 91%	[145]
China (Hebei)	Leaching (Electrochemical decomposition)	−200 mesh, NaOH 40 wt%, 120 °C, slot current density 750 A/m^2, M(alkali:ore) 4:1, 1000 rpm, 6 h	V 95% Cr 90%	[146]
	Roasting	−200 mesh, NaOH-$NaNO_3$ binary melt reaction system, liquid/solid 4:1, base/salt 1:1, 400 °C, 6 h, O_2 flow 0.5 L/min	V 93.7% Cr 88.2%	[155]
China	Roasting	NaOH-added pellet, 700 °C, R(Na/Cr) 7.67, 15 min	V 99.2%	[148]
	Calcification roasting Ammonium carbonate leaching	−74 μm, Calcification roasting: V/Ca molar ratio 1:1.1, muffle furnace, 900 °C, 2 h Ammonium carbonate leaching: $(NH_4)_2CO_3$ 600 g/L, 70 min, 80 °C, S/L 1:20	V 96%	[149]
	Roasting Leaching	−200 mesh, Roasting: No additives, 950 °C, 2.5 h Leaching: Na_2CO_3 160 g/L, 95 °C, 1 h, S/L 1/10 palacios	V 94.13%	[150]
	Salt roasting Leaching Solvent extraction Stripping	Roasting: Sodium salt Water leaching Solvent extraction: LK-N21 15 vol%, pH 5.0, 15 min Stripping: NaOH 0.1 M, 45 °C, 15 min	V > 90% Cr > 90% in raffinate	[151]
N/A	Roasting (Liquid oxidation)	−200 mesh, NaOH-$NaNO_3$ binary molten salt medium, 375 °C, 1 h, 700 rpm, O_2 flow 0.3 L/min, liquid (NaOH-$NaNO_3$)/ore 4:1, M(NaOH/$NaNO_3$) 2.2:1.8	V 90% Cr 80%	[156]
	Roasting (Liquid oxidation)	KOH 75 wt%, 240 °C, 1 h, liquid (KOH mass)/ore 4:1, 700 rpm, O_2 flow 1 L/min	V 95% Cr 93%	[152]

All the roasting methods above showed more than 90% recovery of V, and some of them included a minimum of 80% of Cr. The mixture of sodium chloride and sodium carbonate achieved especially 96% of V and 91% of Cr [145]. In the case of NaOH–$NaNO_3$ binary molten salt medium, the roasting temperature was dropped from 850 °C to 375~400 °C by the traditional method [156]. With potassium hydroxide, Li et al. (2012) tested three reaction media, such as sub molten KOH, sub-molten NaOH, and NaOH–$NaNO_3$ binary molten salt medium. They also dropped the roasting temperature to 200~450 °C [152]. Yan et al. (2015) put no additives for roasting at 950 °C and subsequently conducted soda roasting, which still presented a 94.13% recovery of V [150]. These attempts indicate that it is promising to establish a novel roasting method without toxic emissions.

Even without sodium salt roasting, Wang et al. (2018) established a clean metallurgical process with combined methods of pyro- and hydrometallurgy and produced V with 99.1% purity and Cr with 98.9% purity [138]. Furthermore, there have been efforts to adapt modern technology to vanadium processing. For example, electrochemistry, such as electro-oxidation, has been introduced in leaching. This method can break the silicon layer and expose more V. Hence, there is a possibility to develop further and apply this method to the industry [135,144,146].

It is interesting to note the attempts to extract valuable metals from the residue of slag processing [157–159]. The residue of slag processing is called 'V–Cr-bearing reduced (reducing) residue', and is from wastewater after roasting and neutralisation of vanadium slag [157]. V–Cr-bearing reduced slag contains 5.1% V and 13.4% Cr [158]. The methods for extracting V from V–Cr-bearing reduced residue include selective oxidation and alkaline leaching [157] and a comprehensive approach: leaching with sulphuric acid, selective oxidation, and precipitation of vanadium and chromium [158]. These methods cover separation and purification as well, showing great recoveries of V and Cr of 98.7% and 99.4%, respectively [159]. This case shows the fast development of utilising secondary waste. Therefore, developing processes to recover critical and strategic metals from secondary and tertiary sources should continue regardless of market status or metal price to the material cycle as much as possible.

5.5. Titanium Slag

In this chapter, titanium slag will be discussed briefly. Its chemical compositions are listed in Table 11, and its extraction methods are reviewed below.

Table 11. Compositions (%) of critical and strategic elements in titanium slag.

Sample (Origin)	TiO_2	Ti_2O_3	Al_2O_3	MgO	MnO	V_2O_5	Zn	References
Ti-bearing slag (China)	22.58	-	13.04	6.93	-	-	-	[160]
	16.9	4.06	13.76	8.48	0.53	-	-	[161]
	17.58	3.86	14.08	7.86	-	0.21	-	[162]
	10.4	-	12.5	10.6	-	-	-	[163]
Ti-bearing slag (N/A)	38.2	-	10.4	9.7	0.75	0.92	-	[164]

Titanium slag is produced by a conventional method of titanium production. Ilmenite ore is upgraded to titanium slag and then to synthetic rutile [165]. Accordingly, many publications on titanium slag deal with high grades of titanium, ranging from around 70% to more than 90% as a main product [166–168]. Thus, 'titanium-bearing blasting furnace slag' (also called titanium-bearing slag) as a by-product from vanadium–titanium magnetite ore or titanomagnetite ore is more appropriate in this study [169].

The main difference between titanium slag and titanium-bearing slag is the grade of titanium. The latter shows a lower grade, less than 30%, of TiO_2 in the slags [14] and around 1% of V_2O_5, as shown in Table 11. Much literature focuses on the upgrade and extraction of Ti from titanium-bearing slag, such as the removal of impurities [164], the enrichment of anosovite by adding SiO_2 [160], and various oxidation conditions [161,170,171], as well as novel technology such as microwave treatment, super-gravity separation, and ultrasonic processes [14].

Even though Ti is considered one of the most abundant elements in the crust [165], it is critical and strategic for many applications. Thus, studies on the characteristics, mineralogy, and processing of titanium-bearing slag would be highly beneficial for waste management and process development. Another important critical and strategic element in the titanium-bearing slag is V, and Han et al. [162] presented the flowsheet of two-stage oxidation, alkali leaching and acid leaching, to extract Fe, V, and Ti.

6. Conclusions

This review attempted to present the potential of non-ferrous slag as a viable source of critical and strategic metals needed for high-technology and clean energy applications. Slag is considered waste and is mostly dumped in heaps after smelting. Non-ferrous slag materials such as tin slag, copper slag, nickel slag, vanadium slag, and titanium slag contain significant amounts of critical and strategic elements, making them potential secondary resources for such metals. Numerous studies have been conducted to extract critical and strategic metals from slag. However, most are still at the laboratory scale and are yet to be proven at pilot plant levels. In addition, their commercial viability is yet to be assessed for economic and environmental benefits. Technology development, as applied to resource recovery from slag, involves applications of mineral processing, hydrometallurgical, or pyrometallurgical techniques. As the importance of and the demand for critical and strategic metals continues to rise, it will soon be inevitable to consider extracting valuable metals in waste and by-products, such as slag, to augment the global demand for supplies. Extracting values from waste is categorised under technospheric mining, a concept that promotes a circular economy in the mining and resource industries. Until then, developing green and sustainable technologies for metal extraction and filling the knowledge gaps in the recovery of critical and strategic elements from non-ferrous slag should be actively pursued and promoted.

Author Contributions: Conceptualisation, B.L. and R.D.A.; investigation, B.L. and R.D.A.; writing—original draft preparation, B.L.; writing—review and editing, B.L., M.A. and R.D.A.; visualisation, B.L.; supervision, M.A. and R.D.A. All authors have read and agreed to the published version of the manuscript.

Funding: The authors are grateful for the financial support of the Science Industry PhD Fellowship from the Department of Jobs, Tourism, Science, and Innovation (JTSI), Government of Western Australia and Curtin University.

Data Availability Statement: Not applicable.

Acknowledgments: This manuscript is a part of the first author's PhD project.

Conflicts of Interest: Author Mark Aylmore was employed by the company Mineral Resources Ltd. The remaining authors declare that the research was conducted in the absence of any commercial or financial relationships that could be construed as a potential conflict of interest.

References

1. Brandt, D.A.; Warner, J.C. *Metallurgy Fundamentals Ferrous and Nonferrous*, 5th ed.; The Goodheart-Willcox Company, Inc.: Tinley Park, IL, USA, 2009.
2. Piatak, N.M.; Parsons, M.B.; Seal, R.R. Characteristics and environmental aspects of slag: A review. *Appl. Geochem.* **2015**, *57*, 236–266. [CrossRef]
3. Lee, J.-C.; Kurniawan; Kim, E.-Y.; Chung, K.W.; Kim, R.; Jeon, H.-S. A review on the metallurgical recycling of vanadium from slags: Towards a sustainable vanadium production. *J. Mater. Res. Technol.* **2021**, *12*, 343–364. [CrossRef]
4. Dehaine, Q.; Tijsseling, L.T.; Glass, H.J.; Törmänen, T., Butcher, A.R. Geometallurgy of cobalt ores: A review. *Miner. Eng.* **2021**, *160*, 106656. [CrossRef]
5. Zhang, C.; Hu, B.; Wang, H.; Wang, M.; Wang, X. Recovery of valuable metals from copper slag. *Min. Metall. Explor.* **2020**, *37*, 1241–1251. [CrossRef]
6. Shen, H.; Forssberg, E. An overview of recovery of metals from slags. *Waste Manag.* **2003**, *23*, 933–949. [CrossRef] [PubMed]
7. Gbor, P.K.; Mokri, V.; Jia, C.Q. Characterization of smelter slags. *J. Environ. Sci. Health Part A* **2000**, *35*, 147–167. [CrossRef]
8. Gorai, B.; Jana, R.K.; Premchand. Characteristics and utilisation of copper slag—A review. *Resour. Conserv. Recycl.* **2003**, *39*, 299–313. [CrossRef]
9. Tian, H.; Guo, Z.; Pan, J.; Zhu, D.; Yang, C.; Xue, Y.; Li, S.; Wang, D. Comprehensive review on metallurgical recycling and cleaning of copper slag. *Resour. Conserv. Recycl.* **2021**, *168*, 105366. [CrossRef]
10. Lee, J.-C.; Pandey, B.D. Bio-processing of solid wastes and secondary resources for metal extraction—A review. *Waste Manag.* **2012**, *32*, 3–18. [CrossRef]
11. Srichandan, H.; Mohapatra, R.K.; Parhi, P.K.; Mishra, S. Bioleaching approach for extraction of metal values from secondary solid wastes: A critical review. *Hydrometallurgy* **2019**, *189*, 105122. [CrossRef]
12. Potysz, A.; van Hullebusch, E.D.; Kierczak, J.; Grybos, M.; Lens, P.N.L.; Guibaud, G. Copper Metallurgical Slags—Current Knowledge and Fate: A Review. *Crit. Rev. Environ. Sci. Technol.* **2015**, *45*, 2424–2488. [CrossRef]
13. Moskalyk, R.; Alfantazi, A. Processing of vanadium: A review. *Miner. Eng.* **2003**, *16*, 793–805. [CrossRef]
14. Shi, J.; Qiu, Y.; Yu, B.; Xie, X.; Dong, J.; Hou, C.; Li, J.; Liu, C. Titanium Extraction from Titania-Bearing Blast Furnace Slag: A Review. *J. Miner. Met. Mater. Soc.* **2022**, *74*, 654–667. [CrossRef]
15. Liu, H.; Liu, H.; Nie, C.; Zhang, J.; Steenari, B.-M.; Ekberg, C. Comprehensive treatments of tungsten slags in China: A critical review. *J. Environ. Manag.* **2020**, *270*, 110927. [CrossRef]
16. Beylot, A.; Dewulf, J.; Greffe, T.; Muller, S.; Blengini, G.-A. Mineral resources depletion, dissipation and accessibility in LCA: A critical analysis. *Int. J. Life Cycle Assess.* **2024**, *29*, 890–908. [CrossRef]
17. Johansson, N.; Krook, J.; Eklund, M.; Berglund, B. An integrated review of concepts and initiatives for mining the technosphere: Towards a new taxonomy. *J. Clean. Prod.* **2013**, *55*, 35–44. [CrossRef]
18. Palm, V.; Östlund, C. Lead and Zinc Flows from Technosphere to Biosphere in a City Region. *Sci. Total Environ.* **1996**, *192*, 95–109. [CrossRef]

19. Hofstetter, P. *Perspectives in Life Cycle Impact Assessment: A Structured Approach to Combine Models of the Technosphere, Ecosphere, and Valuesphere*; Springer Science & Business Media: Berlin/Heidelberg, Germany, 1998.
20. Karlsson, S. Closing the Technospheric Flows of Toxic Metals: Modeling Lead Losses from a Lead-Acid Battery System for Sweden. *J. Ind. Ecol.* **1999**, *3*, 23–40. [CrossRef]
21. Palm, V. Material Flow Analysis in Technosphere and Biosphere. Ph.D. Thesis, Royal Institute of Technology, Universitetsservice US AB, Stockholm, Sweden, 2002.
22. Zalasiewicz, J.; Williams, M.; Waters, C.N.; Barnosky, A.D.; Palmesino, J.; Rönnskog, A.-S.; Edgeworth, M.; Neal, C.; Cearreta, A.; Ellis, E.C.; et al. Scale and diversity of the physical technosphere: A geological perspective. *Anthr. Rev.* **2017**, *4*, 9–22. [CrossRef]
23. Krook, J.; Svensson, N.; Eklund, M. Landfill Mining: A Critical Review of Two Decades of Research. *Waste Manag.* **2011**, *32*, 513–520. [CrossRef]
24. Krook, J.; Carlsson, A.; Eklund, M.; Frändegård, P.; Svensson, N. Urban mining: Hibernating copper stocks in local power grids. *J. Clean. Prod.* **2011**, *19*, 1052–1056. [CrossRef]
25. Reid, S.; Tam, J.; Yang, M.; Azimi, G. Technospheric Mining of Rare Earth Elements from Bauxite Residue (Red Mud): Process Optimization, Kinetic Investigation, and Microwave Pretreatment. *Sci. Rep.* **2017**, *7*, 15252. [CrossRef] [PubMed]
26. Sapsford, D.; Cleall, P.; Harbottle, M. In Situ Resource Recovery from Waste Repositories: Exploring the Potential for Mobilization and Capture of Metals from Anthropogenic Ores. *J. Sustain. Metall.* **2017**, *3*, 375–392. [CrossRef]
27. Kim, J.; Azimi, G. Technospheric mining of niobium and titanium from electric arc furnace slag. *Hydrometallurgy* **2020**, *191*, 105203. [CrossRef]
28. Lim, B.; Alorro, R.D. Technospheric Mining of Mine Wastes: A Review of Applications and Challenges. *Sustain. Chem.* **2021**, *2*, 686–706. [CrossRef]
29. Pramanik, S.; Kumari, A.; Sahu, S.K.; Munshi, B. Extraction of metal values from iron-rich mine tailings via chloridized roasting and water leaching. *Waste Manag. Bull.* **2024**, *2*, 113–121. [CrossRef]
30. Simandl, G.J.; Akam, C.; Paradis, S. Which materials are 'critical' and 'strategic'. In *Symposium on Strategic and Critical Metals*; British Columbia Ministry of Energy and Mines: Victoria, BC, Canada, 2015; pp. 1–4.
31. Jin, Y.; Kim, J.; Guillaume, B. Review of critical material studies. *Resour. Conserv. Recycl.* **2016**, *113*, 77–87. [CrossRef]
32. Hofmann, M.; Hofmann, H.; Hagelüken, C.; Hool, A. Critical raw materials: A perspective from the materials science community. *Sustain. Mater. Technol.* **2018**, *17*, e00074. [CrossRef]
33. Ailshie, J.F. Minerals in National Defense. *ABA Sec. Miner. Nat. Res. L. Proc.* **1939**, *70*. Available online: https://heinonline.org/HOL/LandingPage?handle=hein.journals/pabminnl&div=6&id=&page= (accessed on 3 July 2024).
34. Antwerpen, F.V. Strategic Raw Materials. *Ind. Eng. Chem.* **1939**, *31*, 520–530. [CrossRef]
35. Gunn, G. *Critical Metals Handbook*; Wiley: Hoboken, NJ, USA, 2013; pp. 1–439.
36. Korinek, J.; Kim, J. Export restrictions on strategic raw materials and their impact on trade and global supply. *J. World Trade* **2011**, *45*, 255–281.
37. Cote, G. Hydrometallurgy of Strategic Metals. *Solvent Extr. Ion Exch.* **2000**, *18*, 703–727. [CrossRef]
38. Bortnikov, N.S.; Lobanov, K.V.; Volkov, A.V.; Galyamov, A.L.; Vikent'ev, I.V.; Tarasov, N.N.; Distler, V.V.; Lalomov, A.V.; Aristov, V.V.; Murashov, K.Y.; et al. Strategic metal deposits of the Arctic Zone. *Geol. Ore Depos.* **2015**, *57*, 433–453. [CrossRef]
39. European Commission. *Study on the Critical Raw Materials for the EU 2023—Final Report*; Publications Office ofthe European Union: Luxembourg, 2023.
40. Australian Government. *Critical Minerals List 2023 Update*; Critical Minerals Office: Canberra, ACT, Australia, 2023.
41. Su, Y.; Hu, D. Global Dynamics and Reflections on Critical Minerals. *E3S Web Conf.* **2022**, *352*, 03045. [CrossRef]
42. Applegate, J.D. *2021 Draft List of Critical Minerals*; U.S. Geological Survey, Department of the Interior, Ed.; Federal Publisher: Alexandria, NSW, Australia, 2021; pp. 62199–62203.
43. BGS. *Risk List*; British Geological Survey: Nottingham, UK, 2015.
44. Mines, M.O. *Critical Minerals For India*; Government of India: New Delhi, India, 2023.
45. OECD. *OECD Secretary-General's Report to Ministers 2021*; OECD Publishing: Paris, France, 2021.
46. OECD. *A Sustainable Materials Management Case Study Critical Metals and Mobile Devices*; OECD Publishing: Paris, France, 2011.
47. USGS. *2022 Final List of Critical Minerals*; Department of the Interior: Washington, DC, USA, 2022.
48. United States Department of Energy. *Critical Minerals and Materials*; United States Department of Energy: Washington, DC, USA, 2021.
49. Suliva, J.; Deese, B. *Building Resilient Supply Chains, Revitalizing American Manufacturing, and Fostering Broad-Based Growth*; The White House: Washington, DC, USA, 2021; pp. 1–250.
50. Nassar, N.T.; Fortier, S.M. *Methodology and Technical Input for the 2021 Review and Revision of the US Critical Minerals List*; U.S. Geological Survey: Reston, VA, USA, 2021.
51. Anes, I.A.; Garjulli, F.; Siqueira de Carvalho, M.; Tenório, J.A.S.; Espinosa, D.C.R.; Coleti, J.L. Extraction of niobium in one step from tin slag by NH_4F-HCl leaching process. *Can. J. Chem. Eng.* **2024**, *102*, 168–176. [CrossRef]
52. Brocchi, E.A.; Moura, F.J. Chlorination methods applied to recover refractory metals from tin slags. *Miner. Eng.* **2008**, *21*, 150–156. [CrossRef]
53. Gaballah, I.; Allain, E.; Djona, M. Extraction of tantalum and niobium from tin slags by chlorination and carbochlorination. *Metall. Mater. Trans. B* **1997**, *28*, 359–369. [CrossRef]

54. Gaballah, I.; Allain, E. Recycling of strategic metals from industrial slag by a hydro-and pyrometallurgical process. *Resour. Conserv. Recycl.* **1994**, *10*, 75–85. [CrossRef]
55. Odo, J.U.; Okafor, W.C.; Ekpe, S.O.; Nwogbu, C.C. Extraction of Niobium from Tin Slag. *Int. J. Sci. Anc Res. Publ.* **2014**, *4*. Available online: http://www.ijsrp.org/research-paper-1114.php?rp=P353322 (accessed on 3 July 2024).
56. Subramanian, C.; Suri, A.K. Recovery of niobium and tantalum from low grade tin slag—A hydrometallurgical approach. *Environ. Waste Manag. Non-Ferr. Metall. Ind.* **1998**, *831*, 100–107.
57. Allain, E.; Kanari, N.; Diot, F.; Yvon, J. Development of a process for the concentration of the strategic tantalum and niobium oxides from tin slags. *Miner. Eng.* **2019**, *134*, 97–103. [CrossRef]
58. Sulaiman, M.Y. Simultaneous determination of thorium and uranium in tin slag. *Sci. Total Environ.* **1993**, *130–131*, 187–195. [CrossRef]
59. Chirikure, S.; Heimann, R.B.; Killick, D. The technology of tin smelting in the Rooiberg Valley, Limpopo Province, South Africa, ca. 1650–1850 CE. *J. Archaeol. Sci.* **2010**, *37*, 1656–1669. [CrossRef]
60. Gaballah, I.; Allain, E.; Meyer-Joly, M.C.; Malua, K. A possible method for the characterization of amorphous slags: Recovery of refractory metal oxides from tin slags. *Metall. Mater. Trans. B* **1992**, *23*, 249–259. [CrossRef]
61. Gustison, R. *Electric Furnace Method of Beneficiating Tantalum-And Noibium-Containing Tin Slags and The Like*; Kawecki Berylco Industries, Inc.: New York, NY, USA, 1973.
62. Soedarsono, J.W.; Permana, S.; Hutauruk, J.K.; Adhyputra, R.; Rustandi, A.; Maksum, A.; Widana, K.S.; Trinopiawan, K.; Anggraini, M. Upgrading tantalum and niobium oxides content in Bangka tin slag with double leaching. *IOP Conf. Ser. Mater. Sci. Eng.* **2018**, *316*, 012052. [CrossRef]
63. Munir, B.; Permana, B.; Amilia, A.; Maksum, A.; Soedarsono, J.W. Initial Study for Cerium and Lanthanum Extraction from Bangka Tin Slag through NaOH and $HClO_4$ Leaching. *MATEC Web Conf.* **2019**, *269*, 07003. [CrossRef]
64. Tylecote, F.; Photos-Jones, E.; Earl, B. The composition of tin slags from the south-west of England. *World Archaeol.* **1989**, *20*, 434–445. [CrossRef]
65. Mudzanapabwe, N.T.; Chinyamakobvu, O.S.; Simbi, D.J. In situ carbothermic reduction of a ferro-columbite concentrate in the recovery of Nb and Ta as metal matrix composite from tin smelting slag waste dump. *Mater. Des.* **2004**, *25*, 297–302. [CrossRef]
66. Fredericci, C.; Pizani, P.S.; Morelli, M.R. Crystallization of blast furnace slag glass melted in SnO_2 crucible. *J. Non-Cryst. Solids* **2007**, *353*, 4062–4065. [CrossRef]
67. Permana, S.; Soedarsono, J.W.; Rustandi, A.; Maksum, A. Other Oxides Pre-removed from Bangka Tin Slag to Produce a High Grade Tantalum and Niobium Oxides Concentrate. *IOP Conf. Ser. Mater. Sci. Eng.* **2016**, *131*, 012006. [CrossRef]
68. Dhir, R.K.; de Brito, J.; Silva, R.V.; Lye, C.Q. 8—Use of Glass Cullet in Road Pavement Applications, in Sustainable Construction Materials. In *Sustainable Construction Materials: Glass Cullet*; Obe, R.K.D., De Brito, J., Ghataora, G.S., Lye, C.Q., Eds.; Woodhead Publishing: Sawston, UK, 2018; pp. 297–325.
69. Dhir, R.K.; de Brito, J.; Silva, R.V.; Lye, C.Q. 9—Use of Glass Cullet in Ceramics and Other Applications, in Sustainable Construction Materials. In *Sustainable Construction Materials: Glass Cullet*; Obe, R.K.D., De Brito, J., Ghataora, G.S., Lye, C.Q., Eds.; Woodhead Publishing: Sawston, UK, 2018; pp. 327–387.
70. Wang, L.; Xu, Y.; Tian, L.; Chen, Y.; Yang, A.; Gan, G.; Ma, Y.; Du, Y. Two-stage method recovery of metals from copper slag: Realize the resource utilization of all components. *Miner. Eng.* **2024**, *206*, 108503. [CrossRef]
71. Miganei, L.; Gock, E.; Achimovičová, M.; Koch, L.; Zobel, H.; Kähler, J. New residue-free processing of copper slag from smelter. *J. Clean. Prod.* **2017**, *164*, 534–542. [CrossRef]
72. Teğin, İ.; Ziyadanogullari, R. Recovery of Copper and Cobalt from Converter Slag with a New Flotation Method Using H2S. *J. Environ. Sci. Eng. A* **2018**, *7*, 69–75.
73. Yuksel, U.; Teğin, İ.; Ziyadanogullari, R. Recovery of Copper and Cobalt from Copper Slags as Selective. *J. Environ. Sci. Eng. A* **2017**, *6*, 388–394. [CrossRef]
74. Bulut, G.; Perek, K.T.; Gűl, A.; Arslan, F.; Őnal, G. Recovery of metal values from copper slags by flotation and roasting with pyrite. *Min. Metall. Explor.* **2007**, *24*, 13–18. [CrossRef]
75. Bulut, G. Recovery of copper and cobalt from ancient slag. *Waste Manag. Res.* **2006**, *24*, 118–124. [CrossRef]
76. Arslan, C.; Arslan, F. Recovery of copper, cobalt, and zinc from copper smelter and converter slags. *Hydrometallurgy* **2002**, *67*, 1–7. [CrossRef]
77. Kıyak, B.; Őzer, A.; Altundoğan, H.S.; Erdem, M.; Tűmen, F. Cr(VI) reduction in aqueous solutions by using copper smelter slag. *Waste Manag.* **1999**, *19*, 333–338. [CrossRef]
78. Altundoğan, H.S.; Tümen, F. Metal recovery from copper converter slag by roasting with ferric sulphate. *Hydrometallurgy* **1997**, *44*, 261–267. [CrossRef]
79. Yucel, O.; Addemir, O.; Tekin, A.; Nizamoglu, S. Recovery of Cobalt from Copper Slags. *Miner. Process. Extr. Metall. Rev.* **1992**, *10*, 99–107. [CrossRef]
80. Mikoda, B.; Potysz, A.; Kmiecik, E. Bacterial leaching of critical metal values from Polish copper metallurgical slags using Acidithiobacillus thiooxidans. *J. Environ. Manag.* **2019**, *236*, 436–445. [CrossRef] [PubMed]
81. Potysz, A.; Kierczak, J.; Fuchs, Y.; Grybos, M.; Guibaud, G.; Lens, P.N.L.; van Hullebusch, E.D. Characterization and pH-dependent leaching behaviour of historical and modern copper slags. *J. Geochem. Explor.* **2016**, *160*, 1–15. [CrossRef]

82. Potysz, A.; Kierczak, J.; Pietranik, A.; Kądziołka, K. Mineralogical, geochemical, and leaching study of historical Cu-slags issued from processing of the Zechstein formation (Old Copper Basin, southwestern Poland). *Appl. Geochem.* **2018**, *98*, 22–35. [CrossRef]
83. Rudnik, E.; Burzyńska, L.; Gumowska, W. Hydrometallurgical recovery of copper and cobalt from reduction-roasted copper converter slag. *Miner. Eng.* **2009**, *22*, 88–95. [CrossRef]
84. Yang, W.J.; Wang, C.Y.; Ma, B.Z.; Wang, Z.; Chen, Y.Q.; Yin, F. Recovery of cobalt from copper converter slag by a selective reduction-roasting process. *Adv. Mater. Res.* **2012**, *550*, 2186–2189. [CrossRef]
85. Zhai, X.-J.; Li, N.-J.; Zhang, X.; Fu, Y.; Jiang, L. Recovery of cobalt from converter slag of Chambishi Copper Smelter using reduction smelting process. *Trans. Nonferrous Met. Soc. China* **2011**, *21*, 2117–2121. [CrossRef]
86. Yang, Z.; Rui-lin, M.; Wang-dong, N.; Hui, W. Selective leaching of base metals from copper smelter slag. *Hydrometallurgy* **2010**, *103*, 25–29. [CrossRef]
87. Banza, A.N.; Gock, E.; Kongolo, K. Base metals recovery from copper smelter slag by oxidising leaching and solvent extraction. *Hydrometallurgy* **2002**, *67*, 63–69. [CrossRef]
88. Tshiongo, N.; Mbaya, R.K.K.; Maweja, K.; Tshabalala, L.C. Effect of cooling rate on base metals recovery from copper matte smelting slags. *World Acad. Sci. Eng. Technol.* **2010**, *46*, 273–277.
89. Vítková, M.; Ettler, V.; Johan, Z.; Kríbek, B.; Sebek, O.; Mihaljevic, M. Primary and secondary phases in copper-cobalt smelting slags from the Copperbelt Province, Zambia. *Mineral. Mag.* **2010**, *74*, 581–600. [CrossRef]
90. Ettler, V.; Johan, Z.; Kříbek, B.; Šebek, O.; Mihaljevič, M. Mineralogy and environmental stability of slags from the Tsumeb smelter, Namibia. *Appl. Geochem.* **2009**, *24*, 1–15. [CrossRef]
91. Rozendaal, A.; Horn, R. Textural, mineralogical and chemical characteristics of copper reverb furnace smelter slag of the Okiep Copper District, South Africa. *Miner. Eng.* **2013**, *52*, 184–190. [CrossRef]
92. Kaksonen, A.H.; Särkijärvi, S.; Puhakka, J.A.; Peuraniemi, E.; Junnikkala, S.; Tuovinen, O.H. Solid phase changes in chemically and biologically leached copper smelter slag. *Miner. Eng.* **2017**, *106*, 97–101. [CrossRef]
93. Palacios, J.; Sánchez, M. Wastes as resources: Update on recovery of valuable metals from copper slags. *Miner. Process. Extr. Metall.* **2011**, *120*, 218–223. [CrossRef]
94. Parada, F.; Parra, R.; Watanabe, T.; Hino, M.; Palacios, J.; Sánchez, M. Recovery of iron-molybdenum alloy from copper slags. In Proceedings of the VIII International Conference on 'Molten Slags, Fluxes and Salts', Santiago, Chile, 18–21 January 2009.
95. Cardona Valencia, N. The Physical Chemistry of Copper Smelting Slags and Copper Losses at the Paipote Smelter Part 2—Characterisation of industrial slags. *Can. Metall. Q.* **2011**, *50*, 330–340. [CrossRef]
96. Sukla, L.B.; Panda, S.C.; Jena, P.K. Recovery of cobalt, nickel and copper from converter slag through roasting with ammonium sulphate and sulphuric acid. *Hydrometallurgy* **1986**, *16*, 153–165. [CrossRef]
97. Anand, S.; Rao, K.S.; Jena, P.K. Pressure leaching of copper converter slag using dilute sulphuric acid for the extraction of cobalt, nickel and copper values. *Hydrometallurgy* **1983**, *10*, 305–312. [CrossRef]
98. Anand, S.; Rao, P.K.; Jena, P.K. Recovery of metal values from copper converter and smelter slags by ferric chloride leaching. *Hydrometallurgy* **1980**, *5*, 355–365. [CrossRef]
99. Manasse, A.; Mellini, M.; Viti, C. The copper slags of the Capattoli Valley, Campiglia Marittima, Italy. *Eur. J. Mineral. Eur. J. Miner.* **2001**, *13*, 949–960. [CrossRef]
100. Michailova, I.; Mehandjiev, D. Characterization of Fayalite from Copper Slags. *J. Univ. Chem. Technol. Metall.* **2010**, *45*, 317–326.
101. Deng, T.; Ling, Y. Processing of copper converter slag for metal reclamation. Part I: Extraction and recovery of copper and cobalt. *Waste Manag. Res. J. Int. Solid Wastes Public Clean. Assoc. ISWA* **2007**, *25*, 440–448. [CrossRef]
102. USGS. *Mineral Commodity Summaries 2024*; U.S. Geological Survey: Reston, VA, USA, 2024; p. 212.
103. LME. *Cobalt Price*; LME: London, UK, 2024.
104. Jones, R.; Denton, R.T.; Reynolds, Q.G.; Parker, J.A.L.; van Tonder, G.J.J. Recovery of cobalt from slag in a DC arc furnace at Chambishi, Zambia. *J. South. Afr. Inst. Min. Metall.* **2002**, *102*, 5–9.
105. Skirrow, R.G.; Huston, D.L.; Mernagh, T.P.; Thorne, J.P.; Dulfer, H.; Senior, A.B. *Critical Commodities for a High-Tech World: Australia's Potential to Supply Global Demand*; Geoscience Australia: Canberra, NSW, Australia, 2013.
106. Ziyadanogullari, B. Recovery of Copper and Cobalt from Concentrate and Converter Slag. *Sep. Sci. Technol.* **2000**, *35*, 1963–1971. [CrossRef]
107. Hamamci, C.; Ziyadanoğullari, B. Effect of Roasting with Ammonium Sulfate and Sulfuric Acid on the Extraction of Copper and Cobalt from Copper Converter Slag. *Sep. Sci. Technol.* **1991**, *26*, 1147–1154. [CrossRef]
108. Tümen, F.; Bailey, N.T. Recovery of metal values from copper smelter slags by roasting with pyrite. *Hydrometallurgy* **1990**, *25*, 317–328. [CrossRef]
109. Arslan, F.; Giray, K.; Őnal, G.; Gürkan, V. Development of a flowsheet for recovering copper and tin from copper refining slags. *Eur. J. Miner. Process. Environ. Prot.* **2002**, *2*, 94–102.
110. Anand, S.; Das, R.P.; Jena, P.K. Reduction-roasting and ferric chloride leaching of copper converter slag for extracting copper, nickel and cobalt values. *Hydrometallurgy* **1981**, *7*, 243–252. [CrossRef]
111. Yin, F.; Xing, P.; Li, Q.; Wang, C.; Wang, Z. Magnetic separation-sulphuric acid leaching of Cu–Co–Fe matte obtained from copper converter slag for recovering Cu and Co. *Hydrometallurgy* **2014**, *149*, 189–194. [CrossRef]
112. Banda, W.; Morgan, N.; Eksteen, J.J. The role of slag modifiers on the selective recovery of cobalt and copper from waste smelter slag. *Miner. Eng.* **2002**, *15*, 899–907. [CrossRef]

113. Altundogan, H.S.; Boyrazli, M.; Tumen, F. A study on the sulphuric acid leaching of copper converter slag in the presence of dichromate. *Miner. Eng.* **2004**, *17*, 465–467. [CrossRef]
114. Erüst, C.; Akcil, A.; Gahan, C.S.; Tuncuk, A.; Deveci, H. Biohydrometallurgy of secondary metal resources: A potential alternative approach for metal recovery. *J. Chem. Technol. Biotechnol.* **2013**, *88*, 2115–2132. [CrossRef]
115. Potysz, A.; van Hullebusch, E.D.; Kierczak, J. Perspectives regarding the use of metallurgical slags as secondary metal resources—A review of bioleaching approaches. *J. Environ. Manag.* **2018**, *219*, 138–152. [CrossRef] [PubMed]
116. Potysz, A.; Kierczak, J. Prospective (bio) leaching of historical copper slags as an alternative to their disposal. *Minerals* **2019**, *9*, 542. [CrossRef]
117. Vestola, E.A.; Kuusenaho, M.K.; Närhi, H.M.; Tuovinen, O.H.; Puhakka, J.A.; Plumb, J.J.; Kaksonen, A.H. Acid bioleaching of solid waste materials from copper, steel and recycling industries. *Hydrometallurgy* **2010**, *103*, 74–79. [CrossRef]
118. Mehta, K.; Pandey, B. Bio-assisted leaching of copper, nickel and cobalt from copper converter slag. *Mater. Trans. JIM* **1999**, *40*, 214–221. [CrossRef]
119. Parada, F.; Sanchez, M.; Ulloa, A.; Carrasco, J.C.; Palacios, J.; Reghezza, A. Recovery of molybdenum from roasted copper slags. *Miner. Process. Extr. Metall.* **2010**, *119*, 171–174. [CrossRef]
120. Ettler, V.; Kvapil, J.; Šebek, O.; Johan, Z.; Mihaljevič, M.; Ratié, G.; Garnier, J.; Quantin, C. Leaching behaviour of slag and fly ash from laterite nickel ore smelting (Niquelândia, Brazil). *Appl. Geochem.* **2016**, *64*, 118–127. [CrossRef]
121. Jia, C.Q.; Xiao, J.Z.; Orr, R.G. Behavior of metals in discard nickel smelter slag upon reacting with sulfuric acid. *J. Environ. Sci. Health Part A* **1999**, *34*, 1013–1034. [CrossRef]
122. Li, Y.; Perederiy, I.; Papangelakis, V.G. Cleaning of waste smelter slags and recovery of valuable metals by pressure oxidative leaching. *J. Hazard. Mater.* **2008**, *152*, 607–615. [CrossRef]
123. Ibiapina, V.F.; Medeiros, M.E.; Afonso, J.C.; dos Santos, I.D. Oxidative Acidic Leaching of Ferronickel Slag for Sustainable Metals Recovery and Liability Reduction. *J. Sustain. Metall.* **2024**, *10*, 587–602. [CrossRef]
124. Li, Y.; Papangelakis, V.G.; Perederiy, I. High pressure oxidative acid leaching of nickel smelter slag: Characterization of feed and residue. *Hydrometallurgy* **2009**, *97*, 185–193. [CrossRef]
125. Huang, F.; Liao, Y.; Zhou, J.; Wang, Y.; Li, H. Selective recovery of valuable metals from nickel converter slag at elevated temperature with sulfuric acid solution. *Sep. Purif. Technol.* **2015**, *156*, 572–581. [CrossRef]
126. Pan, J.; Zheng, G.-L.; Zhu, D.-Q.; Zhou, X.-L. Utilization of nickel slag using selective reduction followed by magnetic separation. *Trans. Nonferrous Met. Soc. China* **2013**, *23*, 3421–3427. [CrossRef]
127. Gbor, P.K.; Ahmed, I.B.; Jia, C.Q. Behaviour of Co and Ni during aqueous sulphur dioxide leaching of nickel smelter slag. *Hydrometallurgy* **2000**, *57*, 13–22. [CrossRef]
128. Kierczak, J.; Neel, C.; Puziewicz, J.; Bril, H. The mineralogy and weathering of slag produced by the smelting of lateritic Ni ores, Szklary, Southwestern Poland. *Can. Mineral.* **2009**, *47*, 557–572. [CrossRef]
129. Baghalha, M.; Papangelakis, V.G.; Curlook, W. Factors affecting the leachability of Ni/Co/Cu slags at high temperature. *Hydrometallurgy* **2007**, *85*, 42–52. [CrossRef]
130. Sakaroglou, M.; Anastassakis, G. Nickel recovery from electric arc furnace slag by magnetic separation. *J. Min. Metall. A Min.* **2017**, *53*, 3–15. [CrossRef]
131. Mubarok, M.; Hanif, L. Cobalt and nickel separation in nitric acid solution by solvent extraction using Cyanex 272 and Versatic 10. *Procedia Chem.* **2016**, *19*, 743–750. [CrossRef]
132. Koladkar, D.; Dhadke, P. Cobalt–Nickel separation: The extraction of cobalt (II) and nickel (II) with Bis (2-ethylhexyl) Phosphinic Acid (pia-8) in Toluene. *Solvent Extr. Ion Exch.* **2001**, *19*, 1059–1071. [CrossRef]
133. Ayanda, O.S.; Adekola, F.A.; Baba, A.A.; Ximba, B.J.; Fatoki, O.S. Application of Cyanex extractant in cobalt/nickel separation process by solvent extraction. *Int. J. Phys. Sci.* **2013**, *8*, 89–97.
134. Liu, S.; Wang, L.; Chen, J.; Ye, L.; Du, J. Research progress of vanadium extraction processes from vanadium slag: A review. *Sep. Purif. Technol.* **2024**, *342*, 127035. [CrossRef]
135. Liu, Z.; Li, Y.; Chen, M.; Nueraijemaiti, A.; Du, J.; Fan, X.; Tao, C.-Y. Enhanced leaching of vanadium slag in acidic solution by electro-oxidation. *Hydrometallurgy* **2016**, *159*, 1–5. [CrossRef]
136. Zhang, J.; Zhang, W.; Zhang, L.; Gu, S. Mechanism of vanadium slag roasting with calcium oxide. *Int. J. Miner. Process.* **2015**, *138*, 20–29. [CrossRef]
137. Gao, F.; Olayiwola, A.U.; Liu, B.; Wang, S.; Du, H.; Li, J.; Wang, X.; Chen, D.; Zhang, Y. Review of vanadium production part I: Primary resources. *Miner. Process. Extr. Metall. Rev.* **2021**, *43*, 466–488. [CrossRef]
138. Wang, X.; Gao, D.; Chen, B.; Meng, Y.; Fu, Z.; Wang, M. A clean metallurgical process for separation and recovery of vanadium and chromium from V-Cr-bearing reducing slag. *Hydrometallurgy* **2018**, *181*, 1–6. [CrossRef]
139. Wang, G.; Lin, M.-M.; Diao, J.; Li, H.-Y.; Xie, B.; Li, G. Novel strategy for green comprehensive utilization of vanadium slag with high-content chromium. *ACS Sustain. Chem. Eng.* **2019**, *7*, 18133–18141. [CrossRef]
140. Zhang, G.; Zhang, T.; Lü, G.; Zhang, Y.; Liu, Y.; Zhang, W. Effects of Microwave Roasting on the Kinetics of Extracting Vanadium from Vanadium Slag. *JOM* **2016**, *68*, 577–584. [CrossRef]
141. Yang, Z.; Li, H.-Y.; Yin, X.-C.; Yan, Z.-M.; Yan, X.-M.; Xie, B. Leaching kinetics of calcification roasted vanadium slag with high CaO content by sulfuric acid. *Int. J. Miner. Process.* **2014**, *133*, 105–111. [CrossRef]

142. Jiang, T.; Wen, J.; Zhou, M.; Xue, X. Phase evolutions, microstructure and reaction mechanism during calcification roasting of high chromium vanadium slag. *J. Alloys Compd.* **2018**, *742*, 402–412. [CrossRef]
143. Ning, P.; Lin, X.; Wang, X.; Cao, H. High-efficient extraction of vanadium and its application in the utilization of the chromium-bearing vanadium slag. *Chem. Eng. J.* **2016**, *301*, 132–138. [CrossRef]
144. Liu, Z.; Nueraihemaiti, A.; Chen, M.; Du, J.; Fan, X.; Tao, C. Hydrometallurgical leaching process intensified by an electric field for converter vanadium slag. *Hydrometallurgy* **2015**, *155*, 56–60. [CrossRef]
145. Wang, H.G.; Wang, M.Y.; Wang, X.W. Leaching behaviour of chromium during vanadium extraction from vanadium slag. *Miner. Process. Extr. Metall.* **2015**, *124*, 127–131. [CrossRef]
146. Wang, Z.; Zheng, S.; Wang, S.; Qin, Y.; Du, H.; Zhang, Y. Electrochemical decomposition of vanadium slag in concentrated NaOH solution. *Hydrometallurgy* **2015**, *151*, 51–55. [CrossRef]
147. Wen, J.; Jiang, T.; Zhou, W.; Gao, H.; Xue, X. A cleaner and efficient process for extraction of vanadium from high chromium vanadium slag: Leaching in $(NH_4)_2SO_4$-H_2SO_4 synergistic system and NH_4^+ recycle. *Sep. Purif. Technol.* **2019**, *216*, 126–135. [CrossRef]
148. Ji, Y.; Shen, S.; Liu, J.; Xue, Y. Cleaner and effective process for extracting vanadium from vanadium slag by using an innovative three-phase roasting reaction. *J. Clean. Prod.* **2017**, *149*, 1068–1078. [CrossRef]
149. Li, H.Y.; Wang, K.; Hua, W.-H.; Yang, Z.; Zhou, W.; Xie, B. Selective leaching of vanadium in calcification-roasted vanadium slag by ammonium carbonate. *Hydrometallurgy* **2016**, *160*, 18–25. [CrossRef]
150. Yan, X.-M.; Xie, B.; Jiang, L.; Guo, H.-P.; Li, H.-Y. Leaching of Vanadium from the Roasted Vanadium Slag with High Calcium Content by Direct Roasting and Soda Leaching. In *Rare Metal Technology 2015*; Neelameggham, N.R., Alam, S., Oosterhof, H., Jha, A., Dreisinger, D., Wang, S., Eds.; Springer International Publishing: Cham, Switzerland, 2016; pp. 209–216.
151. Ning, P.; Lin, X.; Cao, H.; Zhang, Y. Selective extraction and deep separation of V(V) and Cr(VI) in the leaching solution of chromium-bearing vanadium slag with primary amine LK-N21. *Sep. Purif. Technol.* **2014**, *137*, 109–115. [CrossRef]
152. Liu, H.B.; Liu, B.; Li, L.J.; Zheng, S.L.; Du, H.; Wang, S.N.; Chen, D.H.; Qi, J.; Zhang, Y. Novel methods to extract vanadium from vanadium slag by liquid oxidation technology. *Adv. Mater. Res.* **2012**, *396*, 1786–1793. [CrossRef]
153. United States Geological Survey. *Mineral Commodity Summaries 2020*; United States Geological Survey: Reston, VA, USA, 2020; p. 204.
154. Li, X.S.; Xie, B. Extraction of vanadium from high calcium vanadium slag using direct roasting and soda leaching. *Int. J. Miner. Metall. Mater.* **2012**, *19*, 595–601. [CrossRef]
155. Liu, B.; Du, H.; Wang, S.-N.; Zhang, Y.; Zheng, S.-L.; Li, L.-J.; Chen, D.-H. A Novel Method to Extract Vanadium and Chromium from Vanadium Slag using Molten NaOH-$NaNO_3$ Binary System. *AIChE J.* **2013**, *59*, 541–552. [CrossRef]
156. Wang, Z.-H.; Zheng, S.-L.; Wang, S.-N.; Liu, B.; Wang, D.-W.; Du, H.; Zhang, Y. Research and prospect on extraction of vanadium from vanadium slag by liquid oxidation technologies. *Trans. Nonferrous Met. Soc. China* **2014**, *24*, 1273–1288. [CrossRef]
157. Chen, B.; Wang, M.; Huang, S.; Ge, Q.; Wang, X.; Sun, B. Extraction of vanadium from V-Cr bearing reduced residue by selective oxidation combined with alkaline leaching. *Can. Metall. Q.* **2018**, *57*, 434–438. [CrossRef]
158. Wang, M.; Chen, B.; Huang, S.; Wang, X.; Liu, B.; Ge, Q.; Xie, S. A novel technology for vanadium and chromium recovery from V-Cr-bearing reducing slag. *Hydrometallurgy* **2017**, *171*, 116–122. [CrossRef]
159. Chen, B.; Huang, S.; Liu, B.; Ge, Q.; Wang, M.; Wang, X. Separation and recovery of vanadium and chromium from acidic leach solution of V-Cr-bearing reducing slag. *J. Environ. Chem. Eng.* **2017**, *5*, 4702–4706. [CrossRef]
160. Li, J.; Zhang, Z.; Zhang, M.; Guo, M.; Wang, X. The Influence of SiO_2 on the Extraction of Ti Element from Ti-bearing Blast Furnace Slag. *Steel Res. Int.* **2011**, *82*, 607–614. [CrossRef]
161. Zhang, L.; Zhang, L.N.; Wang, M.Y.; Li, G.Q.; Sui, Z.T. Recovery of titanium compounds from molten Ti-bearing blast furnace slag under the dynamic oxidation condition. *Miner. Eng.* **2007**, *20*, 684–693. [CrossRef]
162. Han, J.-Q.; Zhang, J.; Zhang, J.-H.; Chen, X.; Zhang, L.; Tu, G.-F. Recovery of Fe, V, and Ti in modified Ti-bearing blast furnace slag. *Trans. Nonferrous Met. Soc. China* **2022**, *32*, 333–344. [CrossRef]
163. Liu, Y.; Chen, X.; Mao, S.; Xiao, Y.; Li, J. Extraction of Valuable Metals from Titanium-bearing Blast Furnace Slag by Acid Leaching. *J. Wuhan Univ. Technol. Mater. Sci. Ed.* **2024**, *39*, 376–385. [CrossRef]
164. Chen, D.-S.; Zhao, L.-S.; Qi, T.; Hu, G.-P.; Zhao, H.-X.; Li, J.; Wang, L.-N. Desilication from titanium–vanadium slag by alkaline leaching. *Trans. Nonferrous Met. Soc. China* **2013**, *23*, 3076–3082. [CrossRef]
165. Fang, Z.Z.; Middlemas, S.; Guo, J.; Fan, P. A new, energy-efficient chemical pathway for extracting Ti metal from Ti minerals. *J. Am. Chem. Soc.* **2013**, *135*, 18248–18251. [CrossRef]
166. Dong, H.; Jiang, T.; Guo, Y.; Chen, J.; Fan, X. Upgrading a Ti-slag by a roast-leach process. *Hydrometallurgy* **2012**, *113*, 119–121. [CrossRef]
167. Chen, G.; Xiong, K.; Peng, J.; Chen, J. Optimization of combined mechanical activation-roasting parameters of titania slag using response surface methodology. *Adv. Powder Technol.* **2010**, *21*, 331–335. [CrossRef]
168. Guo, C.; Chen, J.; Peng, J.-H.; Wan, R.-D. Green evaluation of microwave-assisted leaching process of high titanium slag on life cycle assessment. *Trans. Nonferrous Met. Soc. China* **2010**, *20*, s198–s204.
169. Cai, Y.; Song, N.; Yang, Y.; Sun, L.; Hu, P.; Wang, J. Recent progress of efficient utilization of titanium-bearing blast furnace slag. *Int. J. Miner. Metall. Mater.* **2022**, *29*, 22–31. [CrossRef]

170. Zhang, L.; Zhang, L.; Wang, M.; Li, G.; Sui, Z. Dynamic oxidation of the Ti-bearing blast furnace slag. *ISIJ Int.* **2006**, *46*, 458–465. [CrossRef]
171. Wang, M.; Li, L.; Zhang, L.; Zhang, L.; Tu, G.; Sui, Z. Effect of oxidization on enrichment behavior of TiO_2 in titanium-bearing slag. *Rare Met.* **2006**, *25*, 106–110. [CrossRef]

MDPI AG
Grosspeteranlage 5
4052 Basel
Switzerland
Tel.: +41 61 683 77 34

Metals Editorial Office
E-mail: metals@mdpi.com
www.mdpi.com/journal/metals